UNIVERSITY OF WYOMING

Introduction to Basic Circuits and the Arduino

Approached from a Physics Perspective

Earl Wood and Dr. Michael Pierce

7/1/2015

Foreword

Before discussing how this lab manual is intended to be used, I want to first describe what the lab manual is not designed to do. This lab, despite its title, is not designed to be a replacement for an electrical engineering circuits course. Electrical engineering courses cover the same material over the course of several semesters allowing them to delve into much greater detail and spend much more time than this manual does for similar subjects. Additionally, the final half of the manual is comprised of microprocessing with the Arduino, while not unconnected to electrical engineering, is typically beyond the reach of a basic circuits course. Finally, the lab manual may cover much of the same material as electrical engineering, but it does so from a physics perspective.

The impetus for this manual came from recognition that the average physics student gains very little practical experience with electronics and microprocessing. The result was that a physics student could derive the equation for the magnetic field inside a solenoid but could not tell you how to use an inductor. Or he could program in C, Python, or IDL but cannot use a microprocessor to sample data from a light sensor. This hole in the students' knowledge limits the future graduate's ability to communicate with an engineer at a future job, diagnose and repair equipment in a laboratory setting, or design and build an instrument for a research project.

In retrospect, the reasons for this technology deficit are not hard to see. Technology in general as well as equipment at universities has grown ever more complex and expensive with time. The price of laboratory equipment is now thousands, if not millions of dollars, students are often told to look but do not touch the hardware. Students never learn to explore for themselves, and are instead, restricted to the computer interface. As a result, a generation of physics students has graduated competent in driving a mouse or operating a GUI, but having no knowledge of how the equipment they are operating works, how to fix it should it stop working, or even how to quantify the precision or possible errors in the data obtained from the equipment.

Recognizing the problem, this lab manual then represents our attempt at a solution aided by several fortunate advances in available technology. This point cannot be stressed enough: much of this lab would have been nearly impossible only a few short years ago. The most obvious of these advances is the Arduino; a cheap microprocessor with many I/O pins programmable in a simplified version of C. Suddenly, sampling data from sensors, data logging on computers, and even complex motor control is all possible to a student within just a few hours of receiving the Arduino. Combine this ease of use with a huge online community continually presenting new ideas, and the Arduino becomes the obvious choice for the second half of the lab manual.

While the Arduino is the core of each project, the Arduino owes much of its popularity to the dramatic decrease in price for common electronic components. The explosion of smart phones, tablets, and other portable computers has resulted in the mass production of simple electronic modules. Pieces like touchscreens, accelerometers, and even GPS systems have become commonplace in such devices, resulting in an increase in availability and a decrease in price. From the start of designing the course to finishing the lab manual, the number of modules, shield, sensors, and the like available for purchase and fully compatible with the Arduino have tripled.

While I am very pleased with the final version of the manual and the resulting classes we have taught around it, there are likely many improvements that could be made. I am continually expanding my own knowledge. As I do, I find topics I want to include as well as improvements in how current topics are discussed. Also, as new technology is added by the market, this manual will undoubtedly need to be revised.

For now, I can only hope this manual helps educate students in ways that prepares them for their careers.

Acknowledgements

I want to express my profound thanks to Dr. Michael Pierce, my advisor, without whom, this manual would never have been written. He is responsible for the original idea for the class as well as the tireless fight to see it created

I also want to thank Dr. Daniel Dale for supporting the development of an untested class. He saw value in this class from the beginning and provided key financial support for the startup and continuation each semester.

I owe the students in the first semester of the class a special thanks for being the guinea pigs, and the students from subsequent classes for providing feedback towards the improvement of the manual.

I want to thank Travis Laurance, our lab coordinator, for providing equipment and tolerating our mess.

Finally, I want to thank Henry Wladkowski who took over as the new TA for the class, giving me the time I needed to make improvements.

Table of Contents

Foreword 2

Acknowledgements 4

Lab 0 - Introduction to Laboratory Equipment 9
    Part I: Oscilloscope 11
    Part II: Digital Multimeter 13
    Part III: Breadboard 16
    Part IV: Summary 18
    Part V: Further Applications 18
    Part VI: PreLab HW 19

Lab 1 – The Nuances of Measuring a Circuit & RC Circuits 20
    Part I: Measuring Voltage Drop 21
    Part II: RC Circuits 24
    Part III: RC Circuits as Filters 27
    Part IV: Summary 28
    Part V: Further Applications 29
    Part VI: Prelab Questions 30

Lab 2 - LRC Circuits and Transformers 31
    Part I: LR Circuits 33
    Part II: LRC Circuits 35
    Part III: Transformers 37
    Part IV: Summary 38
    Part V: Further Applications 39
    Part VIII: Prelab Questions 40

Lab 3 - Introduction to Semiconductors: Diodes 41
    Part I: I-V Curve of Diodes 45
    Part II: I-V Curve of Zener Diodes 47
    Part III: Rectifying an AC Voltage: Half-Wave Rectifier 48
    Part IV: Rectifying an AC Voltage: Full-Wave Rectifier 51
    Part V: Buffered Full-wave Rectifier 53

- Part VI: Voltage Regulator 55
- Part VII: Summary 56
- Part VII: Further Applications 56
- Part VIII: Prelab Questions 59

Lab 4 - Introduction to Transistors 60
- Part I: Common Base, Collector, Emitter 65
- Part II: AC Amplifier 72
- Part III: Transistors as a switch 80
- Part IV: Further Applications 82
- Part V: Prelab Questions 83

Lab 5 - Introduction to Operational Amplifiers 84
- Part I: Comparator 86
- Part II: The Unity Gain Buffer 88
- Part III: The Non-inverting Amplifier 89
- Part IV: The Inverting Amplifier 91
- Part V: Integrating Circuits 92
- Part VI Summary 97
- Part VII Further Applications 98
- Part VIII Prelab Questions 101

Lab 6 - Introduction to the Arduino 102
- Part I: Getting Started by Blinking an LED 105
- Part II: Digital and Analog Input/Output 108
- Part III: Analog to Digital Conversion (ADC) 109
- Part IV: Bonus! Investigating the Smoothness of the Potentiometer 111
- Part V: Bonus! Bonus! Make a Dimmer from the Potentiometer 111
- Part VI: Summary 111

Lab 7 - Arduino Output Control 112
- Part I: 7-Segment Display 114
- Part II: Motor Control 116
- Part III: Relay Control 119
- Part IV: SD Card 121
- Part V: Tones and Speakers 123

Part VI: Bonus! Potentiometer Motor Control 123

Part VI: Bonus! Bonus! Display the Speed of the Motor with LEDs 123

Part VII: Turkey! Processing to make a GUI to control LEDs... 123

Part VI: Summary 124

Lab 8 - A Sensor Buffet for the Arduino 125

Part I: Temperature Sensor.. 127

Part II: Photo-resistor... 128

Part III: Infrared Motion Sensor 130

Part IV: Three-axis Tilt Sensor (Accelerometer) 132

Part V: Ultra-sonic Range Finder.. 133

Part VI: Multiple Sensors (Ultrasonic and Accelerometer) 135

Lab 9 - Reference Voltages and Multiplexers 136

Part I: Operating Reference Voltage 138

Part II: Internal Reference Voltages 139

Part III: External Reference Voltages 140

Part IV: Multiplexer Input 142

Part V: Multiplexer Output 143

Lab 10 – The SPI and I2C bus... 144

Part I: SPI Bus – One Master and One Slave 147

Part II: I2C – Bi-directional Communication... 148

Part III: I2C – Query and Receive Data 150

Part IV: I2C: Connecting to a Sensor with inbuilt I2C 152

Part V: I2C Over Long Distances... 156

Part VI: Software Emulated I2C.. 157

Part VII: Summary 157

Part VIII: Further Applications.. 157

Appendix: List of Possibly Useful Websites 160

Java Applets for Electronics: 160

Math Review:... 160

Electronics Review: 160

Problem Solving Strategies: 160

Arduino Links: 160

- Maker's Electronics Links: 160
- Additional Reference Links: 160
- Electronics Tutorial Links: 161
- Semiconductor Physics Links: 161
- Articles and News Links: 161

Appendix B: List of Arduino Sketches 163

Index 194

Lab 0 - Introduction to Laboratory Equipment

Goals: Learn how to operate and test the limits of the

1. Oscilloscope
2. Function Generator
3. Digital Multimeter
4. Breadboard

List of Equipment and Parts

1. Oscilloscope
2. Function Generator
3. DC Power Supply
4. Digital Multimeter
5. Solderless Breadboard
6. Jumper Wires
7. 9 Capacitors (10pf to 220μf)
8. 3 Banana wires
9. 2 Alligator clips
10. 10 Resistors (1Ω to 1MΩ)
11. 1 Fresh AA Battery
12. 1 Dead AA Battery
13. 1 LED

Introduction

To learn the functions of the equipment we will be using for this course, you will be making basic measurements with the Oscilloscope (O'scope) and Digital Multimeter (DMM) as well as testing the limits of what this equipment can measure.

Kikusui Oscilloscope (model 5020): While the oscilloscope (o-scope) in some ways represents old technology it still remains an essential piece of electronic test equipment. Today, signal analysis is often done with software oscilloscopes thanks to the increasing speed of modern computers. We will use the o-scope to display and visualize waves generated by the function generator (see below) in order to obtain a conceptual understanding of the superposition of waves. The figures below show the basic configuration of an oscilloscope.

As can be seen in Figure 0. 1, electrons are boiled off a cathode element and accelerated through a collimation grid via high voltage to produce a collimated beam of electrons. The beam can be focused before passing through two pairs of deflection plates where the amplified signal voltage is applied to steer the beam and draw on a phosphor screen.

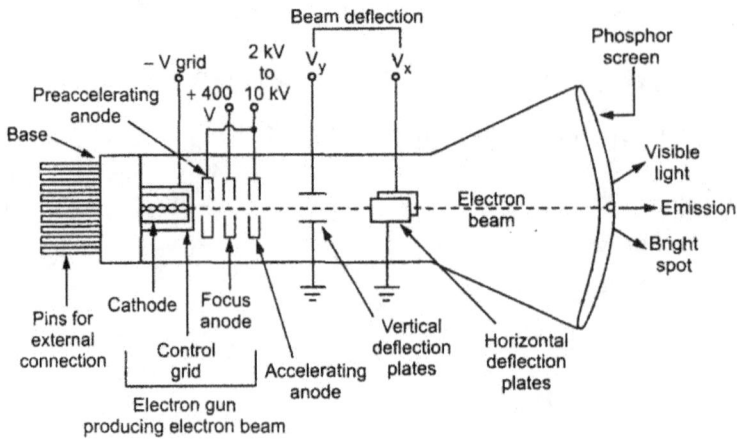

Figure 0. 1 Internal diagram of an Oscilloscope.

Oscilloscope Probes: Oscilloscopes use probes that connect to either channel one or channel two. You can then measure two parts of a circuit simultaneously. The tip has a retractable cowling to protect the conducting wire inside, and shield against accidental contact with any unintended wire (Figure 0. 2). They should never be disassembled, to maintain the contacts inside. They also have a ground clip to complete the circuit for the o-scope across the element of the circuit you are measuring. Finally, a multiplier switch should always be kept at x1.

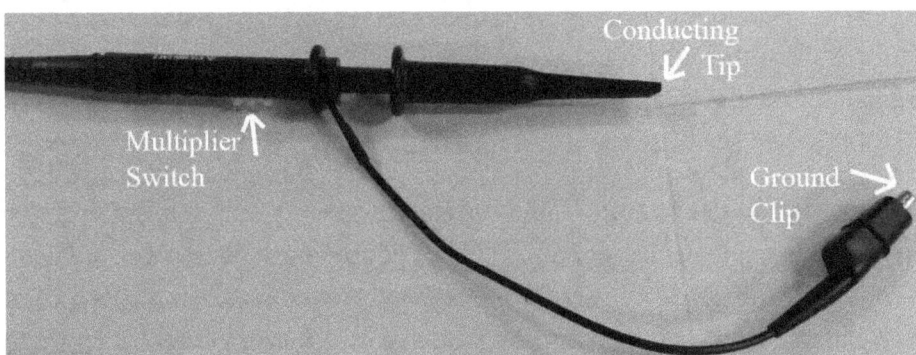

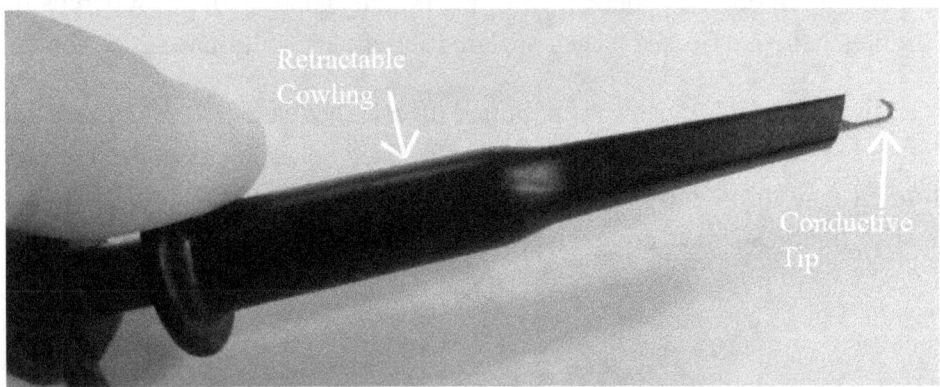

Figure 0. 2: (top) The Oscilloscope Probe. (bottom) A close up of the retractable cowling and conductive tip.

Procedure

Part I: Oscilloscope

- Turn the O'scope on first, and set the "sweep mode" to auto. You should see a solid horizontal line on the screen. Adjust the vertical position and horizontal position knobs until the line is centered on the screen. Then adjust the Illumination, Focus, and Intensity knobs until the line is easily seen and in focus.
- Next connect the O-scope probe to CH1 and connect it to the Function Generator (FG) using a wire clipped in the conducting tip as shown. Make sure:
    1. the "Coupling" is set to AC,
    2. the "Source" is set to Internal,
    3. under "Vertical Mode" push CH1,
    4. under "Position" set to AC,
    5. under "Volts/Div", push in and turn the yellow "VAR" knob fully clockwise until it clicks,
    6. and finally push in and turn the "Variable" knob fully clockwise until it clicks.

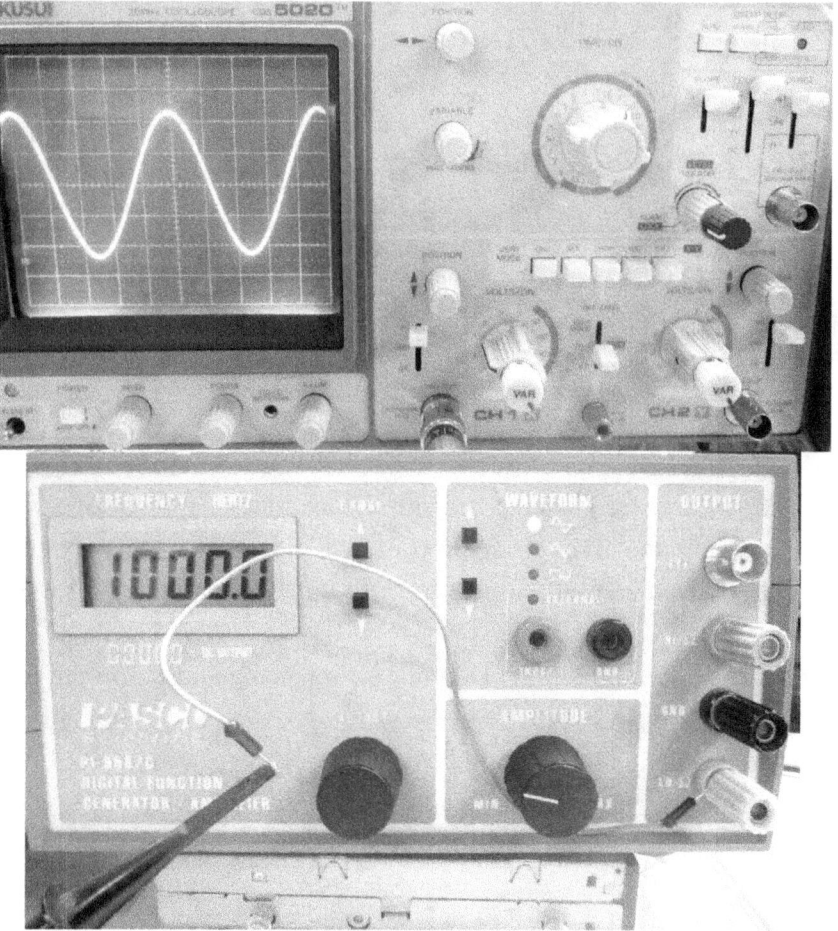

Figure 0. 3: (top) The O-scope showing the output from the FG. (bottom) The FG is connected to the O-scope through the wire clipped in the O-scope probe.

Now, turn on the FG. You should see a sinusoidal graph on the O'scope's display. If it is not steady, adjust the "Level Holdoff" knob until it displays a constant wave on the screen. At this point you are encouraged to play around with the O-scope and FG. The only way you will learn how the instruments work is to change settings and see what happens, so knock yourself out.

All done? Lets continue by making some measurements of the voltage and frequency from the FGs output. The voltage is the amplitude while the wavelength is the period of the wave and the inverse of the frequency. The Volts/Div and Time/Div give you the amount of volts and time for each major grid cell (Figure 0. 4).

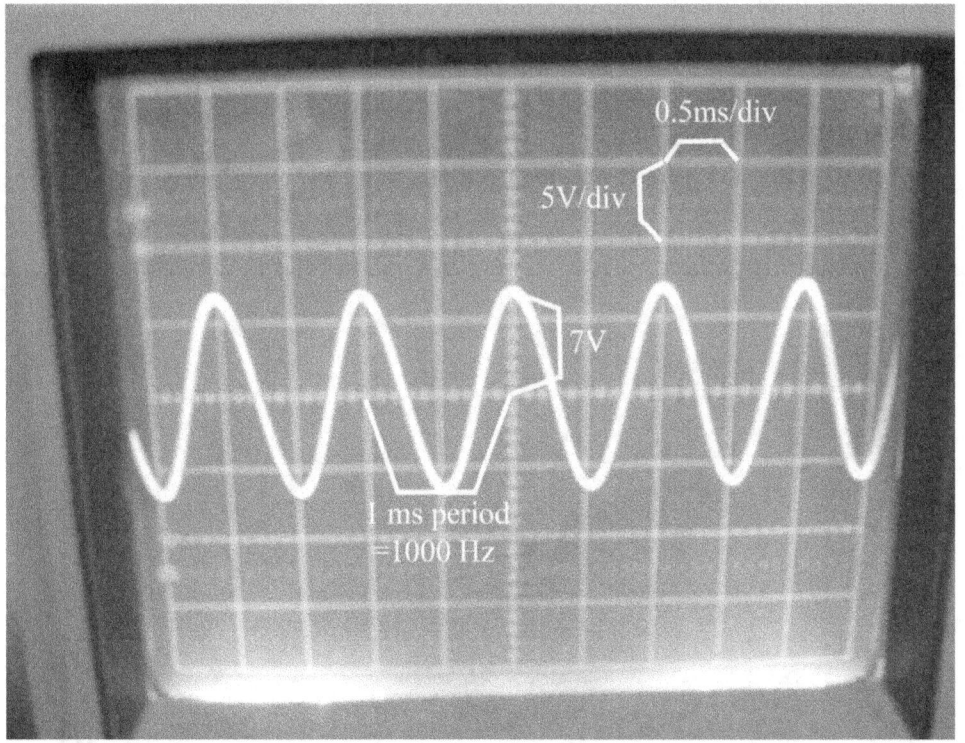

Figure 0. 4: Demonstration of how to measure the amplitude and frequency using an Oscilloscope.

- ➢ Now, choose at least three frequencies and amplitudes that span the range of what the FG can produce and record both the given frequency and the measured frequency and amplitude from the O-scope for each combination along with your estimated error for each measurement.
- ➢ Does the O'scope agree with the FG? Do your errors change based on the frequency or amplitude range?
- ➢ Using this data, write down a good estimate of the error from reading the O'scope for future reference.

Part II: Digital Multimeter

Turn off the O-scope and FG. Next, let's test the Digital Multimeter (DMM).

- If there are no probes (wires) attached, plug a red one into the furthest right socket and a black one in the socket just to the left as shown (Figure 0. 5). Turn the selector to DC voltage.

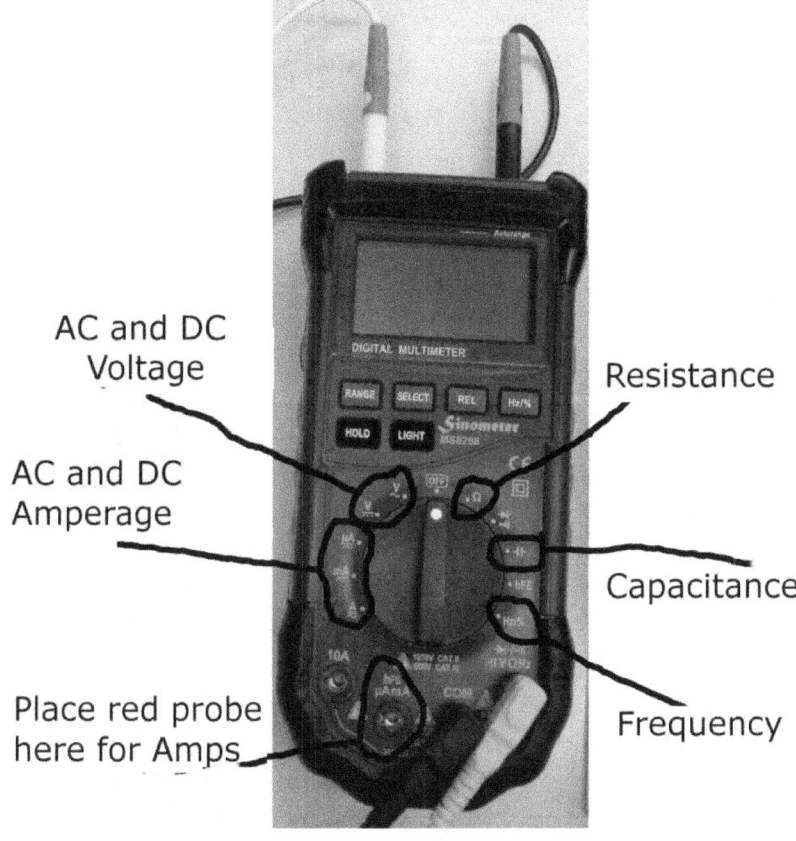

Figure 0. 5: The Multimeter and the important settings.

- You should have a pair of batteries at your station, measure each battery at least three times and record their voltages.
- What is the precision you can expect from the DMM? Does the dead battery read zero volts? These batteries are 1.5 V batteries, do their measurements agree with this? Speculate if not.
- Now, move the red probe to the socket labeled μAmA and turn the selector to mA. Again record the current from shorting each battery like above.
- What is the precision you can expect from the DMM for amperage? What happens if you move the selector to μA; do the measurements agree? Why/Why not?
- In some cases 70 mA across a person's heart is enough to stop your heart. Why does the battery not kill you?

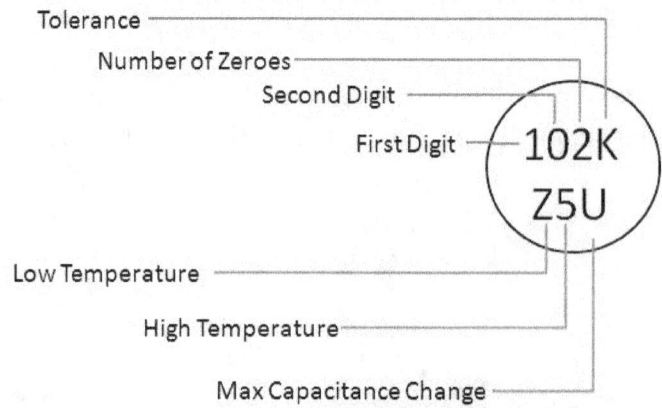

Figure 0. 6: Ceramic Capacitor Codes.

Letter tolerance code		Dielectric codes					
Letter symbol	Tolerance of capacitor	First symbol	Low temperature requirement	Second symbol	High Temperature requirement	Third Symbol	MAX. Capacitance change over temperature
B	+/- 0.10pF	X	-55 ° C	2	+45 ° C	A	+/- 1.0%
C	+/- 0.25pF	Y	-30 ° C	4	+65 ° C	B	+/- 1.5%
D	+/- 0.5pF	Z	+10 ° C	5	+85 ° C	C	+/- 2.2%
E	+/- 0.5%			6	+105 ° C	D	+/- 3.3%
F	+/- 1%			7	+125 ° C	E	+/- 4.7%
G	+/- 2%			8	+150 ° C	F	+/- 7.5%
H	+/- 3%					P	+/- 10.0%
J	+/- 5%					R	+/- 15.0%
K	+/- 10%					S	+/- 22.0%
M	+/- 20%					T	+22%, -33%
N	+/- 30%					U	+22%, -56%
P	+100% ,-0%					V	+22%, -82%
Z	+80%, -20%						

Table 0. 1: (left) Table of Tolerance codes for Ceramic Capacitors. (right) Table of Temperature Coefficients for Ceramic Capacitors.

➢ Move the probes back to their original position and select capacitance.
➢ Find the following capacitors: 10^1pf, 10^2pf, 10^3pf, 10^4pf, 10^5pf ceramic capacitors, and 47μf, 100μf, 150μf, and 220μf electrolytic capacitors.
➢ Measure each capacitor and compare to the stated uncertainty. Notice that the DMM can take several seconds to return a reading. Further notice that some capacitors are unidirectional and show the direction with an arrow or a longer wire.
➢ At some point, you will reach values of capacitance that are smaller or larger than the DMM can measure accurately; what is that value? Do the values agree with the values written on them? What precision can you expect from the capacitors in general? What precision can you expect from the DMM?

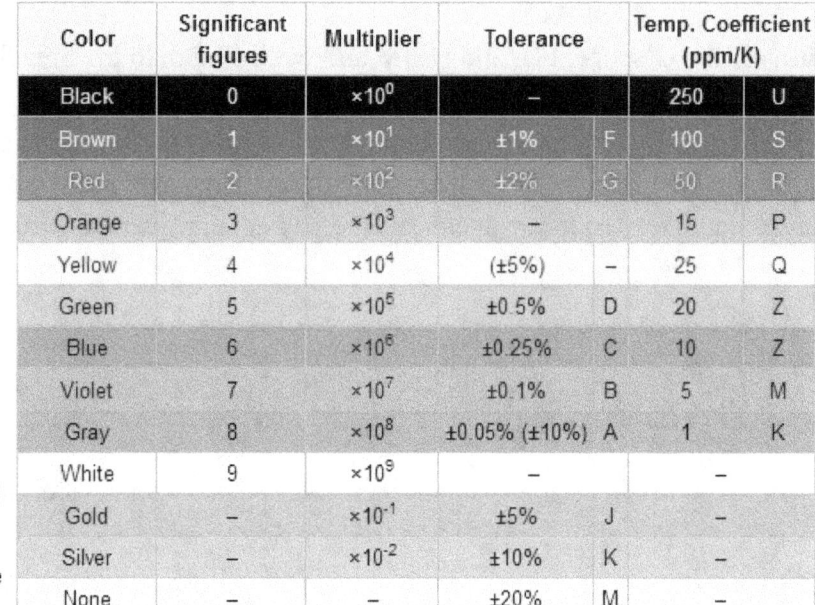
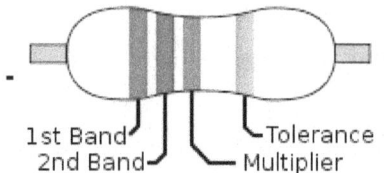

Figure 0. 7: Resistor Color Codes.

Finally, let's measure some resistors with the DMM.

➤ Obtain ten resistors that range in value from 10^0 ohm to 10^6 ohm with two of the resistors possessing a 1% precision referring to the resistor color codes in Figure 0. 7.
➤ Set the DMM to measure Ohms. Measure their resistances and compare to the stated uncertainty.
➤ What is the precision you can expect from the DMM for resistance? Do your measurements match the values listed on each resistor? Why/Why not?

Now to compare the DMM with the O'scope.

➤ Turn on the FG. Choose the same amplitudes and frequencies you chose at the end of Part 1 and measure them with the DMM. Do these value agree with the O'scope? Speculate on why/why not.

Part III: Breadboard

Turn off the O'scope, FG, and DMM and we'll switch to the breadboard. If you have never seen a breadboard before, look at Figure 0. 7 which shows a cutaway and how the wiring is beneath all the holes. As you can see, along the top and bottom, the contacts run the length of the board, while along the middle, the contacts run along the width of the board. This board comes with banana wire jacks so that you can run the power supply to the board.

Figure 0. 8: Cutaway of a bread board to show how electricity is conducted.

➢ Take a jumper wire and connect the jack that you have plugged to the 5V power supply to the board as shown in Figure 0. 9. It is considered good form to run the power supply and grounds to the top and bottom strips of the breadboard.

➢ We want to light an LED that has a maximum forward voltage of 2.2 V. Use the information from the lecture to calculate the values of R1 and R2, and build the voltage divider circuit shown in Figure 0. 9 that will output a voltage of less than 2.2 V across the LED.

➢ Build the Voltage Divider circuit and light the LED. Note that LEDs are unidirectional where the longer contact is positive. If you are correct, once the switch is thrown, the LED should light up. If you are incorrect, the LED will not light and you may have destroyed the LED. Replace the LED and do it right this time!

➢ Test the voltage you calculated using the DMM across both the resistor and the LED.

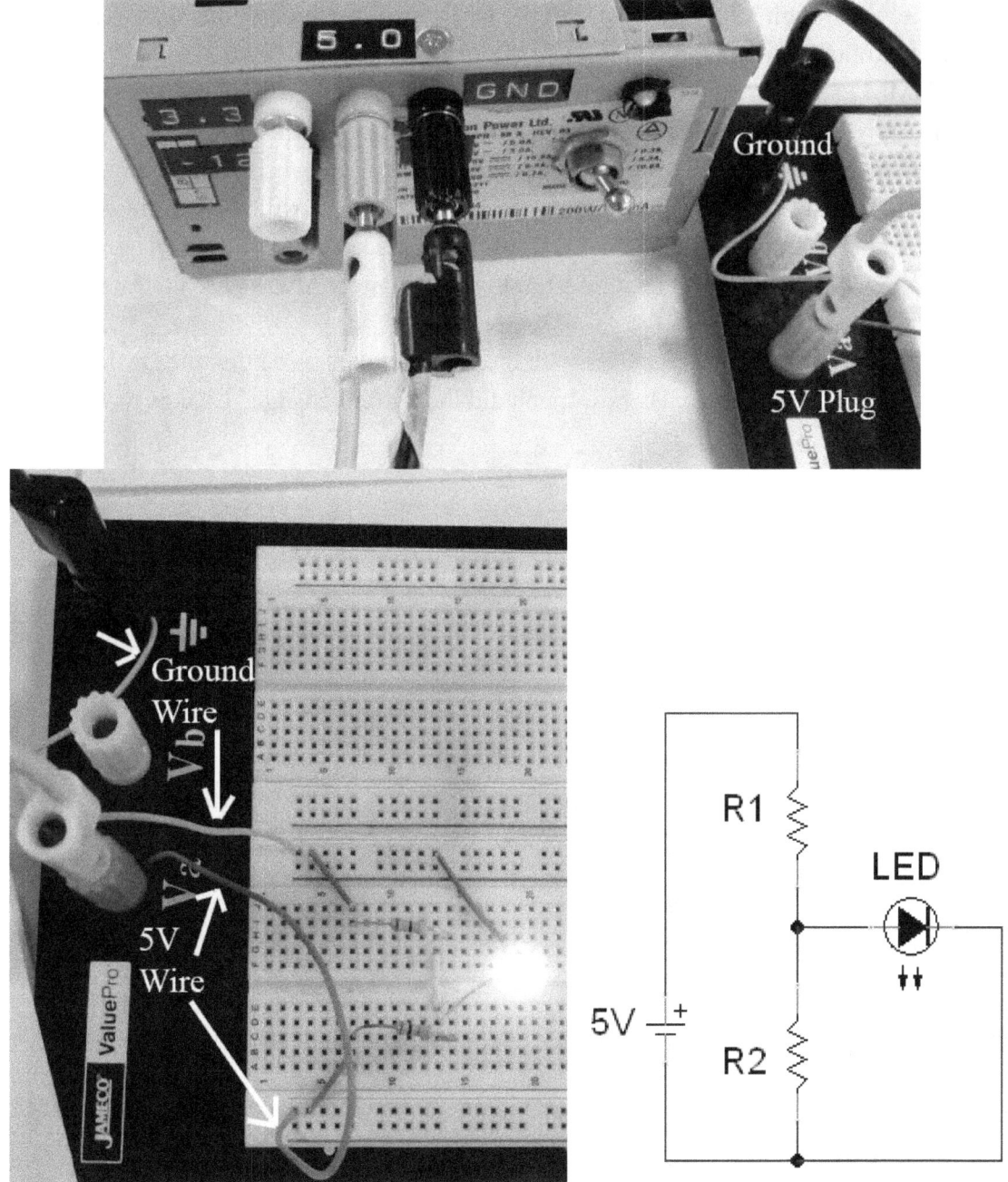

Figure 0. 9: (top) Connecting the power supply to the breadboard. (left) Demonstration of the voltage divider lighting an LED on our breadboard. (right) The circuit diagram for a voltage divider lighting an LED

- Now that you have the LED well lit, we want to measure the current through the circuit. Examine the circuit in Figure 0. 10; you will see that it is the same as the one you currently have built but with the addition of the DMM (shown as circled A).
- Place the DMM in series with the LED as shown and set the DMM to measure current. It is important to remember that current is measured in series, not parallel like the voltage. Currents are different in parallel, but must be the same when in series. This means that you

run the circuit through into the red wire of the DMM and out through the black wire to force the current to flow through the DMM.
- ➢ Set the DMM to measure microAmps and take a reading.
- ➢ The way a DMM measures current is it places a known resistor in series with the rest of the circuit and then measures the voltage across that resistor. It then uses Ohm's law to calculate the current. This means that all attempts to measure the current will always alter the measurement. The smallest setting that you just used, placed the smallest resistor in series with the LED which will result the closest to the real answer. Always use the smallest setting that will give you a result!
- ➢ Test how much of a difference you get with the other settings by setting the DMM to milliAmps and measure again. By how much did the current change? Change to Amps and repeat.

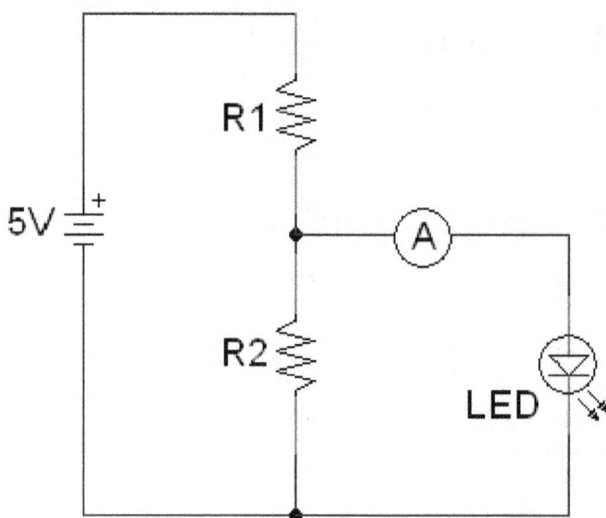

Figure 0. 10: The voltage divider circuit from above to light the LED, but with an Amp-meter in series with the LED.

Part IV: Summary

You should now be familiar with the basics of the main equipment we will be using during the course of the semester. If, during this lab, you feel that you are still uncomfortable with how this equipment works, you should take the time to familiarize yourself because you will need to know it.

Part V: Further Applications

The rest of this course will be filled with uses for this equipment.

Part VI: PreLab HW

1. You are looking at a wave on an oscilloscope. The Volts/Div=2. The Time/Div=.1ms. If there are three divisions between one peak and the nearest trough, what is the Wavelength (or the period) of the wave? The Frequency? If there are 2.5 divisions between the peak and the baseline (zero point), what is the peak-to-peak amplitude?
2. The lab will require error propagation. Derive equation for error of equivalent resistance for 2 resistors in series, and 2 resistors in parallel.
3. For any given two resistors (R1, R2) in series, derive the equation for the value of a third resistor (R3) added in parallel to the other two that will make the equivalent resistance (R_{eq}) half the value of the equivalent resistance for just R1 and R2 in series.

Lab 1 – The Nuances of Measuring a Circuit & RC Circuits

Goals: Investigate the nuances and pitfalls of measuring voltage in simple circuits with the DMM and Oscilloscope as well as RC circuits.

1. Voltage Drop
2. RC Time Constant
3. Impedance
4. Low and High Pass Filtering

List of Equipment and Parts

1. Oscilloscope
2. Function Generator
3. DC Power Supply
4. Digital Multimeter
5. Solderless Breadboard
6. Jumper Wires
7. 4 Capacitors (10pf, 100pf, and two 0.33µf)
8. 5 Resistors (50Ω, 100Ω, three 1kΩ)
9. 3 Banana wires
10. 3 Alligator clips

Introduction: Capacitors

Circuits with capacitors and inductors no longer obey Ohm's law and change their resistance with time. Capacitors slowly build up charge on their plates and will, with enough time, charge to the same voltage as the power supply. At this point, the capacitor acts as an open circuit, not allowing any current to flow.

There are several types of capacitors; the most common are Ceramic, Electrolytic, and Tantalum. Ceramic Capacitors are disk shaped and have no polarity, meaning the order of the contacts is not important. Electrolytic capacitors are cylindrical in shape, and they do have polarity. Polarity can be determined with an arrow, a line at positive, or a longer contact wire at positive. Finally, Tantalum capacitors look much like ceramic ones, but have polarity and generally have much higher precision.

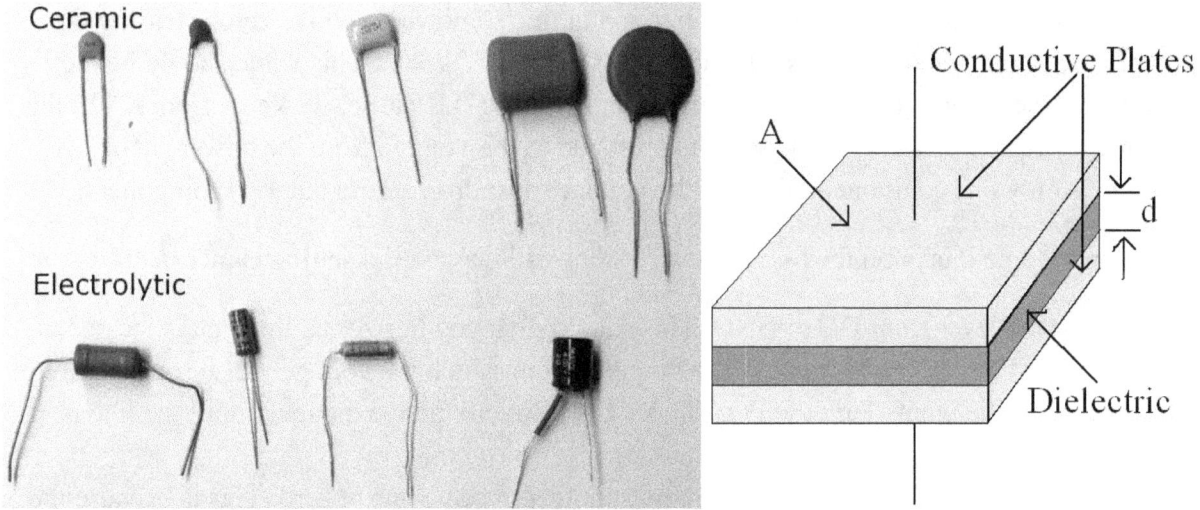

Figure 1. 1: (left(A picture of several different types of capacitors. The ceramic capacitors are non-polar, (right) And a diagram of a simple parallel plate capacitor.

Procedure

Part I: Measuring Voltage Drop

Measuring the voltage drop across individual components in a circuit can be tricky if you are new to oscilloscopes. We will cover both correct methods as well as common mistakes you will likely make in this class.

➢ We'll begin by measuring the voltage drop across a resistor in a simple circuit. Build the circuit in Figure 1. 2 where R is any resistor greater than 1000 Ohms.

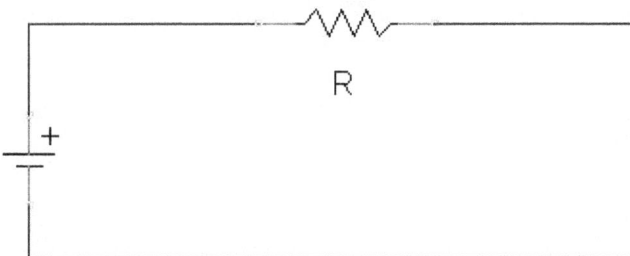

Figure 1. 2: A simple circuit with a 5V DC input and a 1000 Ohm resistor.

➢ Before attaching the o-scope probe, find the switch that has options AC, GND, DC and set it to GND. Then adjust the vertical position knob such that the line on the screen is at Zero.
➢ Switch to the o-scope, and connect the probe across the resistor by connecting the conductive tip to the positive side of the resistor and leave the alligator clip loose. Set the o-scope lever

to measure DC, and you should see a horizontal line somewhere above zero. Measure the voltage of that line using the setting on the VOLTS/DIV knob. This value should be the value of the input voltage from the power supply. This is because, we know from KVL, the voltage drop across all the components must add to the voltage from the power supply. If there is only one component, then all the voltage must drop across that one component.

Lets change some things on the o-scope to see what will happen if a setting is incorrect.

- ➢ First, set the lever from DC to GND. This should give you horizontal line that is located at zero volts. This will always give the zero point from which the voltage will be measured no matter what the input. Turn the POSITION knob directly above the lever until the line is directly on zero on your screen.
- ➢ Next, set the lever from GND to AC. You should still see a line at zero. This is because the AC position tells the o-scope to ignore any DC signal so that the AC signal (such as a sine wave) will be centered about the zero point on the screen. You will use the AC lever soon, but keep in mind that if viewing a DC signal, the AC selection will show only zero.
- ➢ Switch the lever back to DC.
- ➢ During all of this we have left the o-scope probe ground clip loose. It can always be connected to ground in your circuit. Try this by connecting the clip to the back side of the resistor (the side going to ground). This can sometimes help stabilize a noisy signal.
- ➢ The ground clip should not be placed anywhere else in the circuit! To demonstrate why, first turn off the PS. Then set the DMM to measure current. Connect the PS through the DMM to the resistor. The current will now have to flow through the DMM before it reaches the resistor. Turn on the PS and your DMM should read a few milliAmps of current.

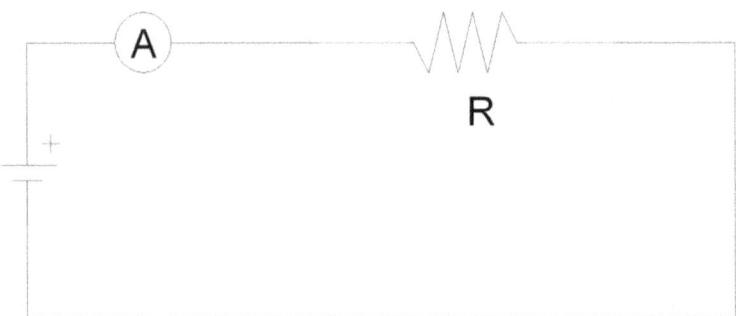

Figure 1. 3: The same circuit as before, but with the DMM in series with the resistor to measure the current.

- ➢ Now, connect the o-scope probe conducting tip to the ground side of the resistor. This next part you must be very careful! BRIEFLY touch the ground clip to the positive side of the resistor while watching the DMM. You should see the current jump.
- ➢ This occurs because the ground clip provides a shortcut for the current to bypass the resistor and go straight to the common ground shared by the o-scope and the PS. If you leave it connected too long, all that power will cause parts of your circuit to melt!

Examine the circuit shown in Figure 1. 4 where R1=R2=R3>1000 Ohms, and note the connection terminals A through F.

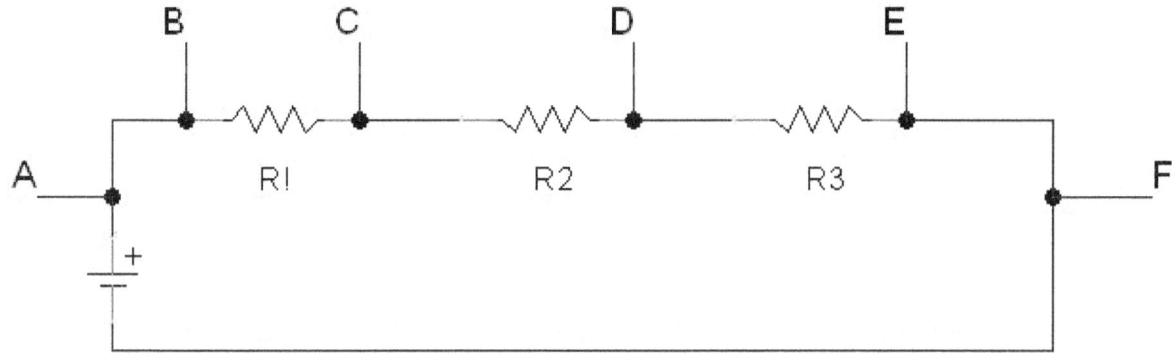

Figure 1. 4: A series resistor circuit. Each letter is a point in the circuit you will measure the voltage.

Because the DMM measures relative voltage, the voltage you measure will depend on where in the circuit you place your probes.

- Build the circuit in Figure 1. 4, and measure across the terminal combinations and record your values in Table 1. 1. Remember to keep the ground clip loose or connected to ground. In your lab report, comment on why the values agree or disagree from DMM to o-scope.

Positive Probe	Neg Probe	O-scope Voltage	DMM Voltage
A	B		
B	C		
B	A		
C	B		
C	D		
B	D		
B	E		
E	F		
E	Ground		
C	Not Used		

Table 1. 1: Measurements to be made of the series resistor circuit with both the o-scope and DMM.

Part II: RC Circuits

Following Kirchhoff's Loop rules, the voltage across a resistor-capacitor pair wired in series must equal the voltage across the capacitor plus the voltage across the resistor. If we write the voltages as a function of time (time domain) then the equation for the voltages in the RC circuit is:

$$v(t) = v_C(t) + v_R(t) \tag{1.1}$$

Using the relations $v_c = Q/C$ and $V_R = I(t)R$, $I(t) = \frac{dq}{dt}$, this equation can be rewritten as:

$$V(t) = \frac{Q(t)}{C} + \frac{dq}{dt}R \tag{1.2}$$

Separating variables, integrating wrt q and t, solving for $Q(t)$, and making use of $C = QV$ yields the voltage across the capacitor as it charges:

$$v_C(t) = V(1 - e^{-t/\tau}) \tag{1.3}$$

This equation is a simplification that assumes the starting voltage, V_i, on the capacitor is zero. Because we will be applying an AC signal to the RC circuit, the voltage across the capacitor will likely be greater than zero. To correct this simplification, we note that V is the voltage difference between the voltage applied to it from the power source (V_{ps}) and the starting voltage on the capacitor (V_i).

$$v_C(t) = (V_{ps} - V_i)(1 - e^{-t/\tau}) \tag{1.4}$$

For discharging, the equation becomes:

$$v_C(t) = V_i(e^{-t/\tau}) \tag{1.5}$$

The voltage across the resistor can then be found by combining Equations (1.1) and (1.3) for charging,

$$v_R(t) = (V_{ps} - V_i)e^{-t/\tau} \tag{1.6}$$

and Equations (1.1) and (1.4) and knowing that $V_{ps} = 0$ for discharging

$$v_R(t) = V_i e^{-t/\tau} \tag{1.7}$$

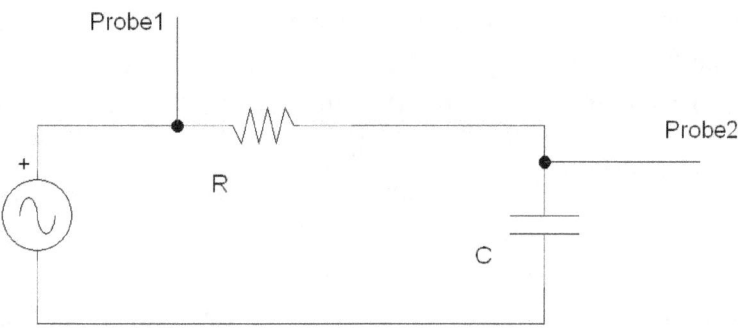

Figure 1. 5: An RC circuit

➢ Choose a capacitor resistor combination that gives a time constant larger than the one you calculated in Prelab Question (3) by roughly a factor of 100. Make sure you choose a capacitor that is not unidirectional.

➢ Measure the capacitor and resistor with the DMM and record their true values within the precision of the DMM
➢ Build the circuit in Figure 1. 5, and use the power from the square wave function generator. You should see on the o-scope a square wave from probe 1 with exponential lines from probe 2 superimposed as seen in Figure 1. 6. If you do not see this, adjust the o-scope until you do.

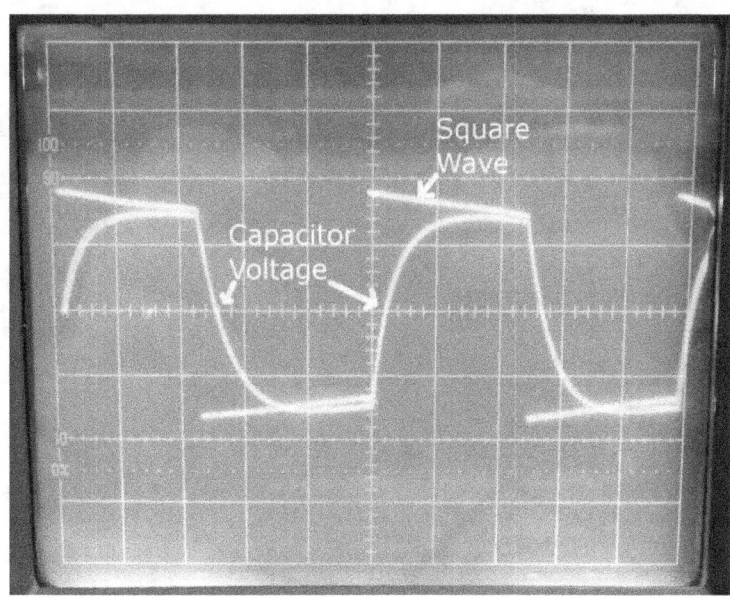

Figure 1. 6: A square wave input to a series RC circuit, along with the voltage across the capacitor. The period of the square wave is long enough, here, to allow the capacitor to fully charge.

➢ The exponential part is the voltage across the capacitor and should be recognizable as the typical RC response from Equation (1. 4). Currently, the period of the wave is large enough

that several time constants will pass before the second half of the waveform applies a negative voltage and starts the charging of the RC circuit over again.
- If we decrease the period of the wave, the RC circuit will be unable to fully charge each time. A less obvious effect is that the RC circuit will be unable to fully discharge as well. This means that the initial voltage across the capacitor will vary based on the frequency.
- Examine Figure 1.7. This is an example of a smaller period wave preventing the voltage across the capacitor from charging completely. We can also see that the voltage rises 63% of the way from the initial voltage to the power supply voltage. This is the value you should have found in Prelab Questions (1).
- Adjust the frequency on the FG until you get your charging voltage across the capacitor to rise 63% of the way from the initial voltage to the power supply voltage. Use Equation (1.4), your work from Prelab Question (1), and the information on the o-scope screen to determine the time constant of your circuit. Compare this value to what you calculated from the resistor and capacitor values.
- Suppose you adjusted the frequency of the FG until the capacitor has only charged to 50% of the value from the initial voltage to the power supply voltage. Set the FG to the frequency you calculated in Prelab Question (2). Adjust the frequency until the voltage is cut at exactly 50% and compare the two values of frequency.

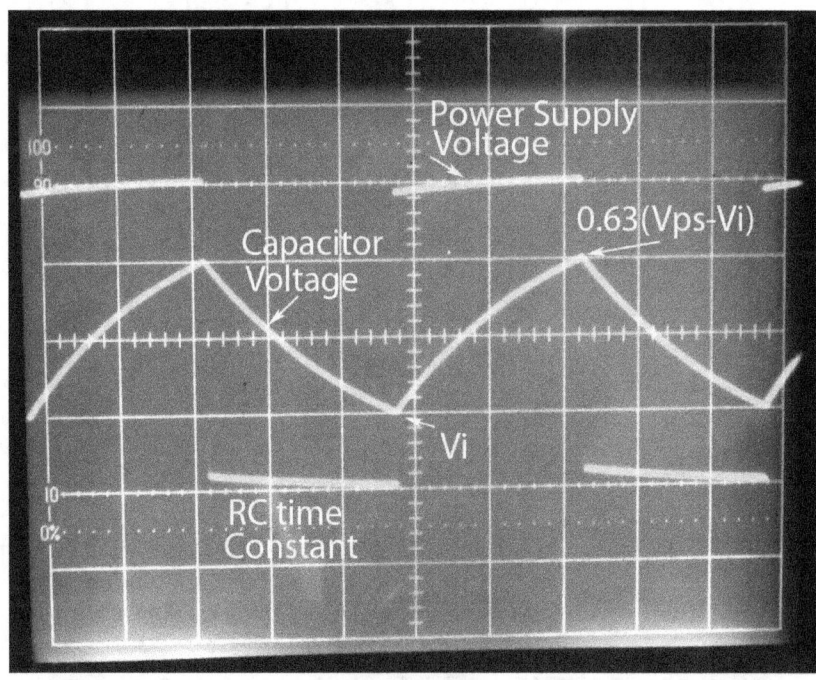

Figure 1.7: A square wave input set to the correct frequency to cut the charge/discharge of the capacitor to just one time constant.

Part III: RC Circuits as Filters

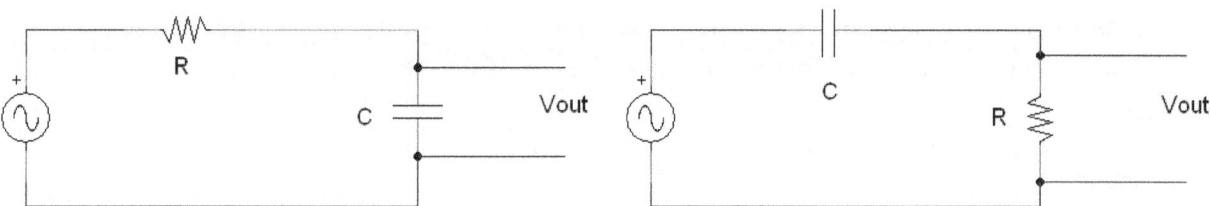

Figure 1. 8: RC series circuit as a filter. (left) Low Pass Filter. (right) High Pass Filter

An RC circuit can be used to filter out unwanted frequencies. When it does so, it is called a "passive filter" because it does not include any amplification to keep the amplitude of the signal from dropping. The RC circuit accomplishes this through the property of Capacitive Reactance, the effective total resistance of the circuit that changes with frequency (Equation (1. 8)).

$$X_c = \frac{1}{2\pi f C} \quad (1.8)$$

Where f is frequency of the AC signal in Hz, and C is the value of the capacitor in farads. We can clearly see that for the correct frequency-capacitor values, the resistance of the circuit would become quite large, acting as an open circuit.

We define the cutoff point for the filter as the point at which the reactance is the same as the resistance from the resistor. If we want to find this frequency, we set Equation (1. 8) equal to R. Solving, we get,

$$f_{cutoff} = \frac{1}{2\pi R C} \quad (1.9)$$

It can also be shown that the output voltage is,

$$V_{out} = V_{in}\left(\frac{X_c}{\sqrt{R^2 + X_c^2}}\right) \quad (1.10)$$

And the phase offset of the output signal in degrees is,

$$\phi = -90 + \tan^{-1}\frac{X_c}{R} \quad (1.11)$$

- Choose a capacitor resistor combination that gives a time constant that you calculated in Prelab Question (4) for a cutoff frequency of about 500 Hz. Make sure you choose a capacitor that is NOT unidirectional.
- Build the low pass filter in Figure 1. 8.
- Input a sine wave from the FG and measure the phase offset and output voltage at the cutoff frequency; compare the output voltage and phase offset to that calculated in Prelab Question (4).

- Investigate the frequency response around the cutoff frequency. Include a few data points lower than the cutoff frequency to show the region where the amplitude does not change. Sample several frequencies before and after the cutoff frequency, recording the output voltage.
- Outside of lab, make a Bode plot by converting your output voltage to decibels and plot verses frequency.
- Build the High Pass filter shown in Figure 1. 8 with a cutoff frequency of 10000Hz.
- Repeat the process above for investigating the voltage around the cutoff frequency for the high pass filter. On the Bode plot, include a few data points higher than the cutoff frequency to show the region where the amplitude does not change.

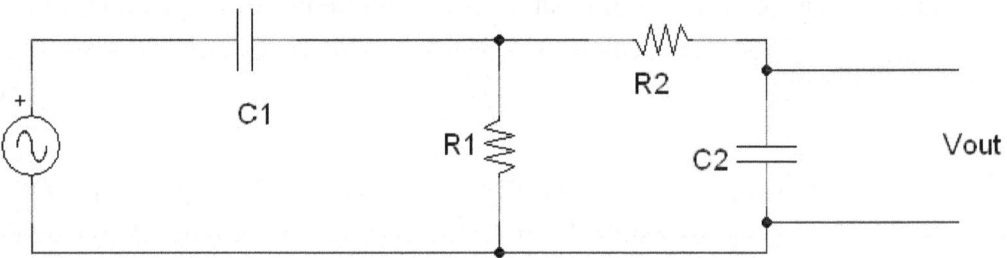

Figure 1. 9: A band pass filter circuit. The high pass filter is on the left, the low pass filter is on the right.

If one wanted to allow only a small range of frequencies, a bandpass filter would be constructed. A bandpass filter is just the combination of the high and low pass filters as shown in Figure 1. 9.

- Build the circuit from the two filters you have just finished, and investigate the frequency response with the same Bode plot you did for the first two passes.

Part IV: Summary

You have now had some exposure to resistors, capacitors, and inductors. The high and low pass filters you built are very common to audio equipment, but usually come in multiple stages as shown in Part VI. Inductors are less frequently used in everyday applications you might come across, but the most common use for inductors we will see is the transformer.

Part V: Further Applications
  i. Nth order High Pass and Low Pass Filter: By adding more stages to a filter, the slopes of the drop-off on the Bode plot become much steeper, decreasing the amount of unwanted frequencies that leak through a single order filter.

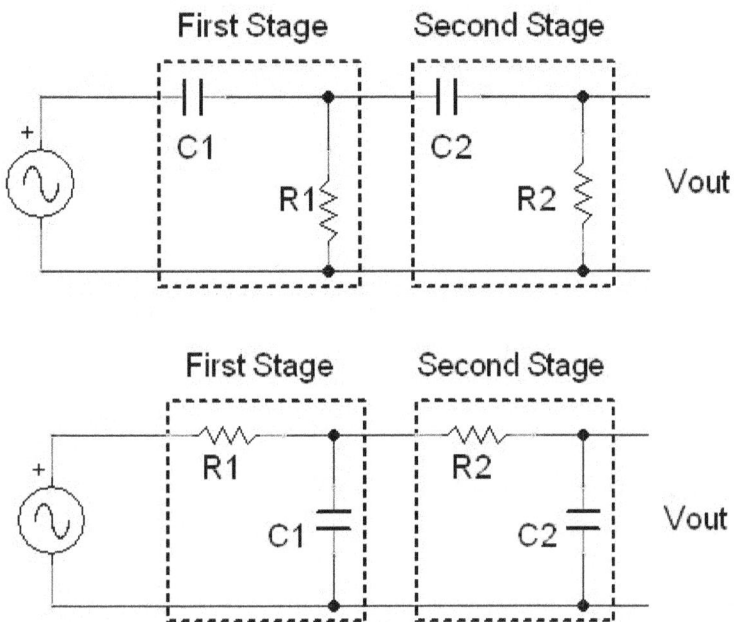

Figure 1. 10:An example of 2nd order filters. 2nd order high pass filter (top). 2nd order low pass filter (bottom).

Descriptions of each of these can be found at this website:

http://www.electronics-tutorials.ws/

Part VI: Prelab Questions

(1) Refer to Part II for how the time constant would be measured. Calculate the voltage across the capacitor as well as the resistor that would result if one time constant ($t = \tau$) has passed. Give your answer in the form of a percentage of the total voltage possible, $\left(\frac{V_C}{V_{ps}-V_i}\right)$. Do this for charging up from some initial voltage, V_i.

(2) Refer to Part II for how the time constant would be measured. Measuring the voltage to the percentage you get in Question (1) can be difficult to tell on the oscilloscope. Measuring a voltage that is half the distance between initial voltage and the power supply voltage could be easier. Calculate the frequency of the input wave that would be necessary to cause the max voltage across the capacitor to reach 50%. Remember that charging time is only half the period of the wave!

(3) Refer to Part II for how the time constant would be measured. Given the maximum time resolution on the oscilloscope as shown by the Time/Div knob (0.2µs), calculate what would be the smallest time constant that could be measured using a square wave from the frequency generator. Assume we want to find the time constant to within a 5% error (half a tick mark) and that the volts reading can be read with no error.

(4) Refer to Part III, you must make both a low and high pass filter. At the cutoff frequency for each filter,
        a. Calculate the time constant.
        b. Calculate the output voltage as a percentage of input voltage.
        c. Calculate the phase shift.

Lab 2 - LRC Circuits and Transformers

Goals: Understand RL, and LRC circuits and:

1. RL Time Constant
2. Damping
3. Resonance
4. Impedance
5. Transformers

List of Equipment and Parts

1. Oscilloscope
2. Function Generator
3. DC Power Supply
4. Digital Multimeter
5. Solderless Breadboard
6. Jumper Wires
7. 1 Capacitor (0.01μf)
8. 2 Resistors (1kΩ, 1.5kΩ)
9. 1 Inductor (800mH)
10. 1 Transformer (5:1 or less)
11. 5 Banana wires
12. 3 Alligator clips

Introduction: L R & C circuits

Inductors are a coil of wire, often around a metal core. They work through the principle of magnetic induction. As the current increases or decreases, a magnetic field is induced that works against the direction of change of the current. So, where a capacitor will eventually block all current in a circuit, the inductor will eventually allow all currents to flow through the circuit. There are two general geometric types of inductors, cylindrical and toroidal. However, the core can be a variety of materials.

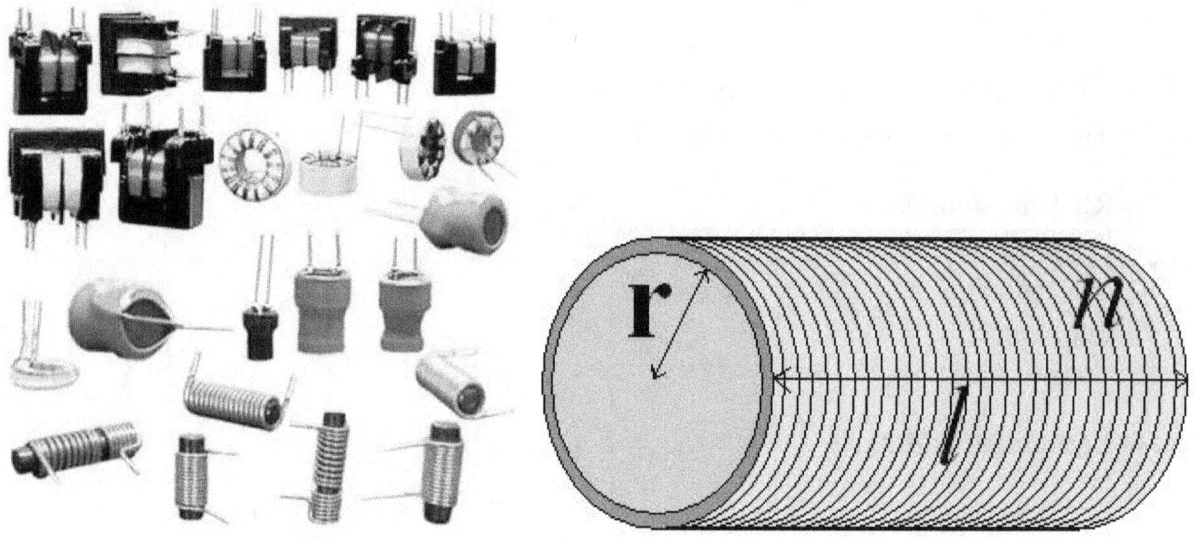

Figure 2. 1: Various types of inductors (left). A functional diagram of an inductor and circuit symbols (right)

Procedure

Part I: LR Circuits

Following Kirchhoff's Loop rules, the voltage across a series wired inductor-resistor pair can be written as:

$$v(t) = v_L(t) + v_R(t) = L\frac{di}{dt} + iR \qquad (2.1)$$

If we follow the same analysis as is done for the RC circuit, we gain equations for the voltage across the inductor and resistor when discharging:

$$V_L(t) = V_i e^{-t/\tau} \qquad (2.2)$$

$$v_R(t) = V_{ps}\left(1 - e^{-\frac{t}{\tau}}\right) \qquad (2.3)$$

Where $\tau = L/R$.

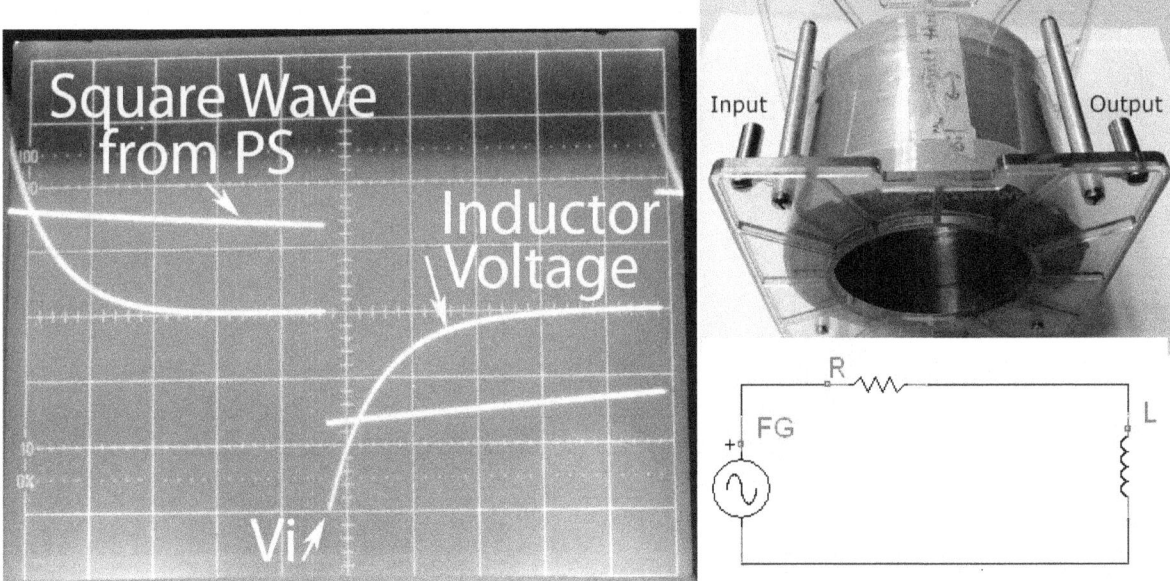

Figure 2.2: (upper right) The inductor coil you will use. (bottom right) An RL series circuit. (left) The voltage output across the inductor with a square wave input.

- Set up the circuit shown in Figure 2.2 using the pictured coil as your inductor and $R \approx 1000 \; ohm$. Connect the o-scope across the inductor to measure the voltage.
- Using a square wave input, adjust the frequency until the voltage across the inductor clearly levels off (Figure 2.2). Notice the spike in the inductor waveform at V_i. Measure the voltage of the spike and explain why you think it exists and has that value compared to the amplitude of the square wave.

- Just as we did with the RC circuit, we will now use the square wave to find the time constant of the circuit. For LR circuits, the voltage across the inductor will decay after spiking. You

LRC Circuits and Transformers

should have determined the percentage of voltage remaining on the inductor after one time constant has passed. Keep that number in mind.

➢ Adjust the frequency of the FG so that the voltage on the inductor decays from V_i to the value you found in Prelab Question (1). Determine the time constant of the circuit from the o-scope.

➢ Determine the value of the inductor in Henrys using your answer from Prelab Question (2).

➢ Now that you have decreased the period and the inductor voltage does not have time to fully discharge, what happened to the value of the spike compared to your previous measurement? Explain why it changed (think about how the inductor stores energy and what happens to the square wave at each spike).

Part II: LRC Circuits

By adding a capacitor to the LR setup, we have essentially added a low pass filter to the high pass LR circuit above. We can attempt to allow a signal to pass through the circuit, but we must balance the reactance from the inductor (high at low frequencies) with the reactance from the capacitor (high at high frequencies). The point where the reactance from both is equal is called the resonant frequency given by,

$$f_r = \frac{1}{2\pi\sqrt{LC}} \qquad (2.4)$$

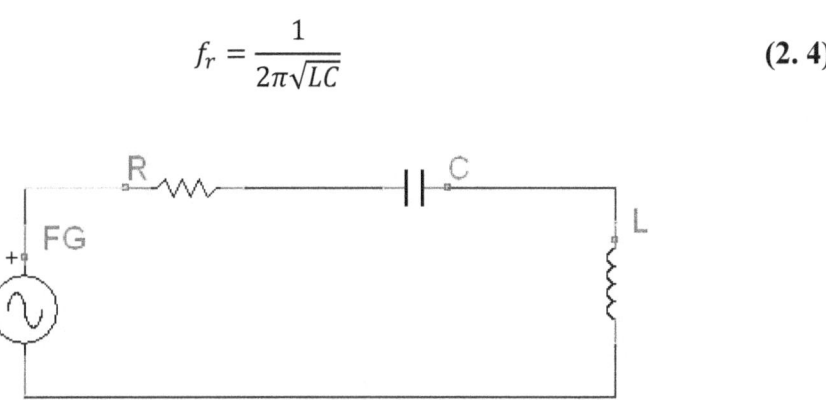

Figure 2.3: An RLC series circuit

- Choose a resistor capacitor combination that will yield a resonant frequency of about 1500 Hz and will filter out frequencies above 8000 Hz if used as an RC low pass filter. Again, measure the resistor and capacitor to keep your errors as low as possible

- Set up the circuit shown in Figure 2.3. First, qualitatively test your low pass filter on a sine wave by attaching the probe across the capacitor. Be certain to use the ground clip on the probe on the back side of the capacitor.
- Switch the probe to measure across the inductor. Sweep through the frequencies surrounding the resonant frequency you calculated. Find the maximum amplitude and record the corresponding frequency. Compare this number to your calculated value. Why might they be different?

- Now, switch the FG to a square wave, and lower the frequency by a factor of 10 compared to the resonant frequency. You should see a sine wave superimposed on the square wave that results from the energy sloshing between the inductor and capacitor. Measure the frequency of this small signal and compare to the other two values of the resonant frequency you have.

➢ Finally, keep the square wave, but increase the frequency by a factor of 10 above your resonant frequency. Observe the waveform that results and compare to the RC circuit output from Part I. Speculate on their similarities or differences.

➢ Make a Bode plot as you did for the RC circuits. Use a sine wave from the FG. How does this band pass filter differ from the one you created from RC circuits?

Part III: Transformers

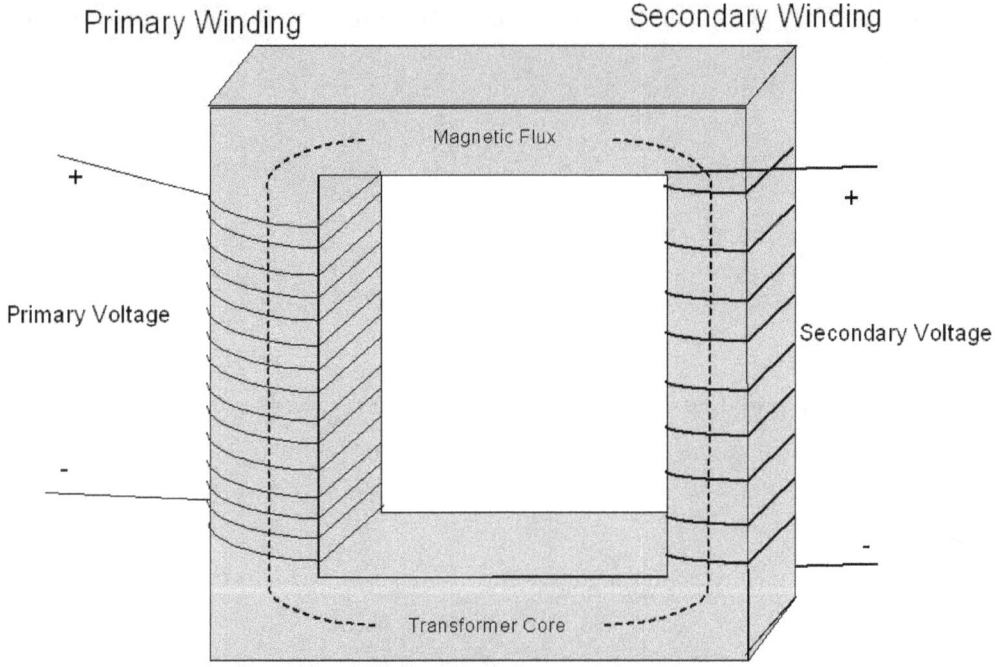

Figure 2. 4: A functional diagram showing how a transformer is constructed.

One extremely important application for inductors is the transformer. As shown in Figure 2. 4, the transformer generally consists of two inductors wound around a common magnetic core. The coil of the primary circuit has N_1 turns and that of the secondary circuit has N_2 turns. The primary coil is connected to an AC voltage source $V_1(t)$ and the secondary coil is connected to a load resistor R_L. In an ideal transformer the core has infinite permeability (μ = infinity), and the magnetic flux is confined within the core. The directions of the currents flowing in the two coils, I_1 and I_2, are defined such that, when I_1 and I_2 are both positive, the flux generated by I_2 is opposite that generated by I_1. Notice that the two inductors, though around the same core, are not physically connected.

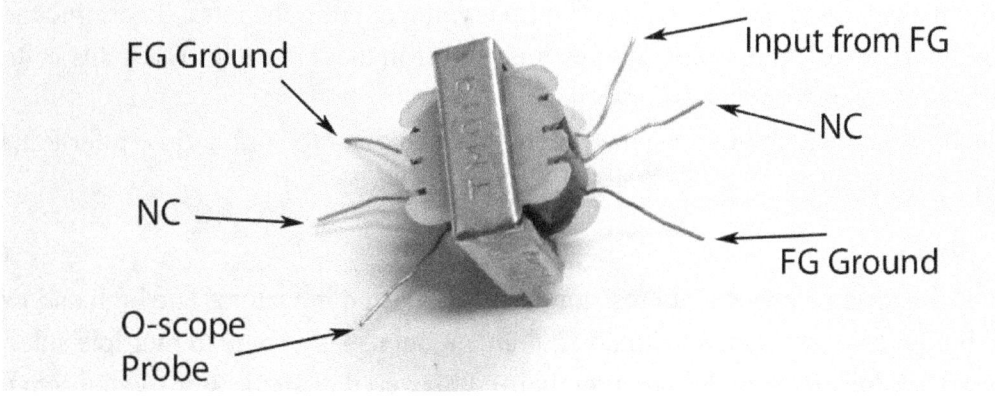

Figure 2. 5: The center tapped transformer you will use in lab.

On the primary side of the transformer, the voltage source V_1 generates a current I_1 in the primary coil, which establishes a flux φ in the magnetic core. The flux φ and the voltage V_1 are related by Faraday's law:

$$V_1 = -N_1 \frac{d\varphi}{dt} \tag{2.5}$$

And, similarly, on the secondary side,

$$V_2 = -N_2 \frac{d\varphi}{dt} \tag{2.6}$$

The combination of which yields:

$$\frac{V_1}{V_2} = \frac{N_1}{N_2} \tag{2.7}$$

Since $P_1 = I_1 V_1$ and $P_2 = I_2 V_2$, then we have:

$$\frac{I_1}{I_2} = \frac{N_1}{N_2} \tag{2.8}$$

> Choose two different colored transformers. If your transformers have three leads on both sides, then ignore the middle lead as it is the center tap (to be used in Lab 3). Connect the FG output wave to the corner on the primary side of the transformer. Then connect the FG ground to the opposite corner leads on both the primary and secondary sides. Then, connect the o-scope probe to the corner opposite the input from the FG and leave the ground clip of the probe unattached.
> Chopping between the input signal from the FG and the output from the secondary side of the transformer, measure the input voltage V_1, and the output voltage V_2 at $1000Hz$. If the input voltage is higher than the output voltage, then you have connected the FG to the secondary coil.
> Measure the ratio between the input and output voltage. From the ratio, determine the turn ratio in the transistor. Do not forget to estimate error in the turn ratio! Does this match what is listed for the transformer you chose?
> What is one major assumption we have made in calculating the turn ratios (refer to the theory section of this lab)?

Part IV: Summary

You have now had some exposure to resistors, capacitors, and inductors. The high and low pass filters you built are very common to audio equipment, but usually come in multiple stages as shown in Part VI. Inductors are less frequently used in everyday applications you might come across, but the most common use for inductors we will see is the transformer.

Part V: Further Applications

i. Tank or Rejector circuit: An LRC circuit in a parallel circuit creates the highest impedance when at the resonant frequency. This property allows the circuit to reject select frequencies when in resonance.

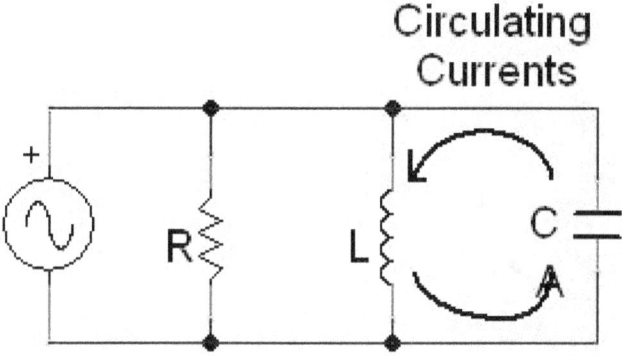

Figure 2. 6: The LRC in parallel rejector circuit.

ii. The Traffic Sensor: Have you ever wondered how the traffic light knows how to change for your car? Well, it is nothing but a big inductor. There is a buried coil of wire just beneath the road, and when your car stops in the center of the coil, your car acts as the inductive core, changing the inductance of the coil. This principle can detect any magnetic metal.

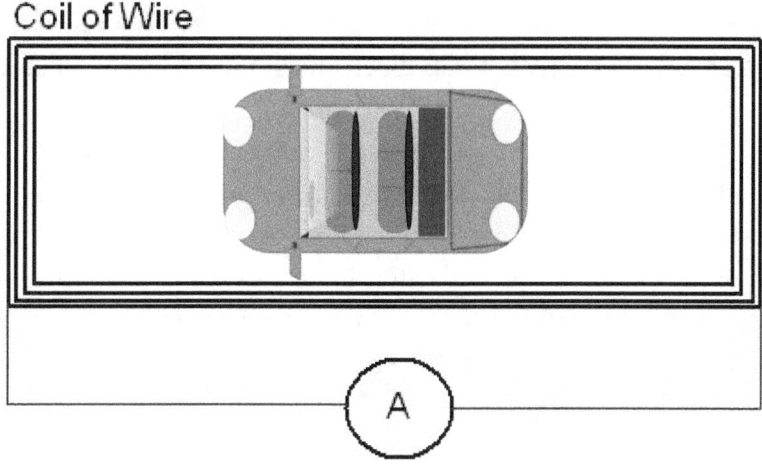

Figure 2. 7: The LC Oscillator and its output voltage.

iii. The Variable Autotransformer, "Variac": This transformer uses a single inductor with a variable tap across any point on the inductor coil. There is no electrical isolation like with a normal transformer, but, when supplied with an AC signal, the Variac can output any fraction of that AC.

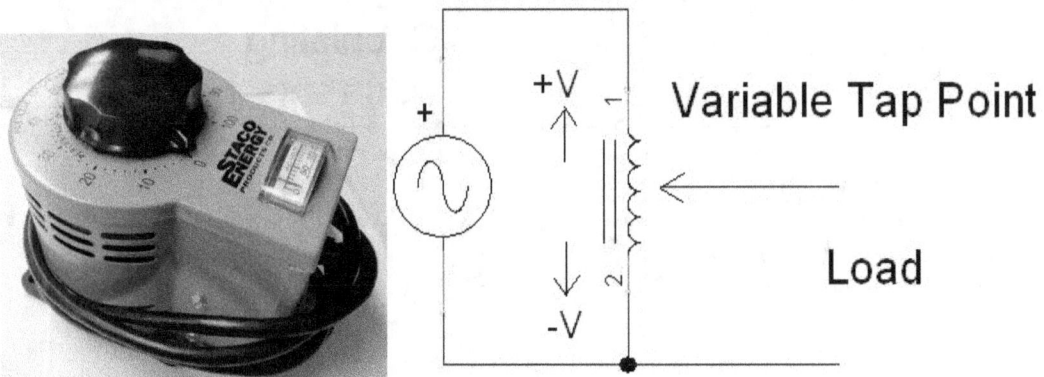

Figure 2. 8: The variac and the circuit diagram.

Descriptions of each of these can be found at this website:

http://www.electronics-tutorials.ws/

Part VIII: Prelab Questions

(1) In Part I, you will have an inductor that decays from some initial voltage, V_i. Calculate the value of the voltage across the inductor as a percentage of the initial voltage after one time constant has passed.

(2) In Part I, you will have to set the frequency to the value needed to clip the voltage decay from the inductor at exactly one time constant. For any given inductor, L, and resistor, R, write an expression for the frequency that would be necessary for this scenario.

(3) Part II asks you to choose a resistor, capacitor, inductor combination that will yield a resonant frequency of about 1500 Hz and will filter out frequencies above 8000 Hz if used as an RC low pass filter. Write two equations, solved for the value of the capacitor and resistor respectively, with the inductor left as the variable 'L'.

Lab 3 - Introduction to Semiconductors: Diodes

Goals: Investigate properties of semiconductor diodes and use them to convert AC to DC. In the process, you will investigate:

1. The Activation and Breakdown voltage of a Diode
2. Halfwave/Fullwave Rectifiers
3. Voltage Ripple

List of Equipment and Parts

1. Oscilloscope
2. Function Generator
3. DC Power Supply
4. 2 Digital Multimeters
5. Solderless Breadboard
6. Jumper Wires
7. 1 Capacitor (10μf)
8. 4 Resistors (100Ω, 200Ω, 1kΩ, 10kΩ)
9. 1 Potentiometer (10kΩ)
10. 4 Diodes (four 1N4001)
11. 1 Zener Diode (1N4733)
12. 1 Transformer (5:1 or less)
13. 4 Banana wires
14. 3 Alligator clips

Introduction and Background

In the previous lab, we used linear components (capacitors, resistors, and inductors). Now we will use the diode, which is a nonlinear device. Diodes make use of the "pn-junction" within a semiconductor. N-type materials have an "excess of electrons" and p-type materials have a "deficit of electrons". Thus, in n-type materials electrons are the charge carriers and in p-type materials "holes" are the charge carriers (Figure 3. 1). When these two materials are brought into contact, they diffuse into each other at the boundary such that the n-type becomes positively charged and the p-type becomes negatively charged. The result is an electric field across the boundary. If an external field is applied, ("forward biasing") the potential decreases allowing charge to can flow across the junction. If the pn-junction is "reversed biased" no or very little current will flow. The result is a diode, which restricts current flow to essentially one direction.

Diodes have a "I-V characteristic" which is the relationship between the current through and the voltage across a circuit component (Figure 3. 1). From this curve, you can also determine the forward turn-on voltage, or voltage drop, V_d of the diode. This is usually about 0.5 volt but varies from diode to diode.

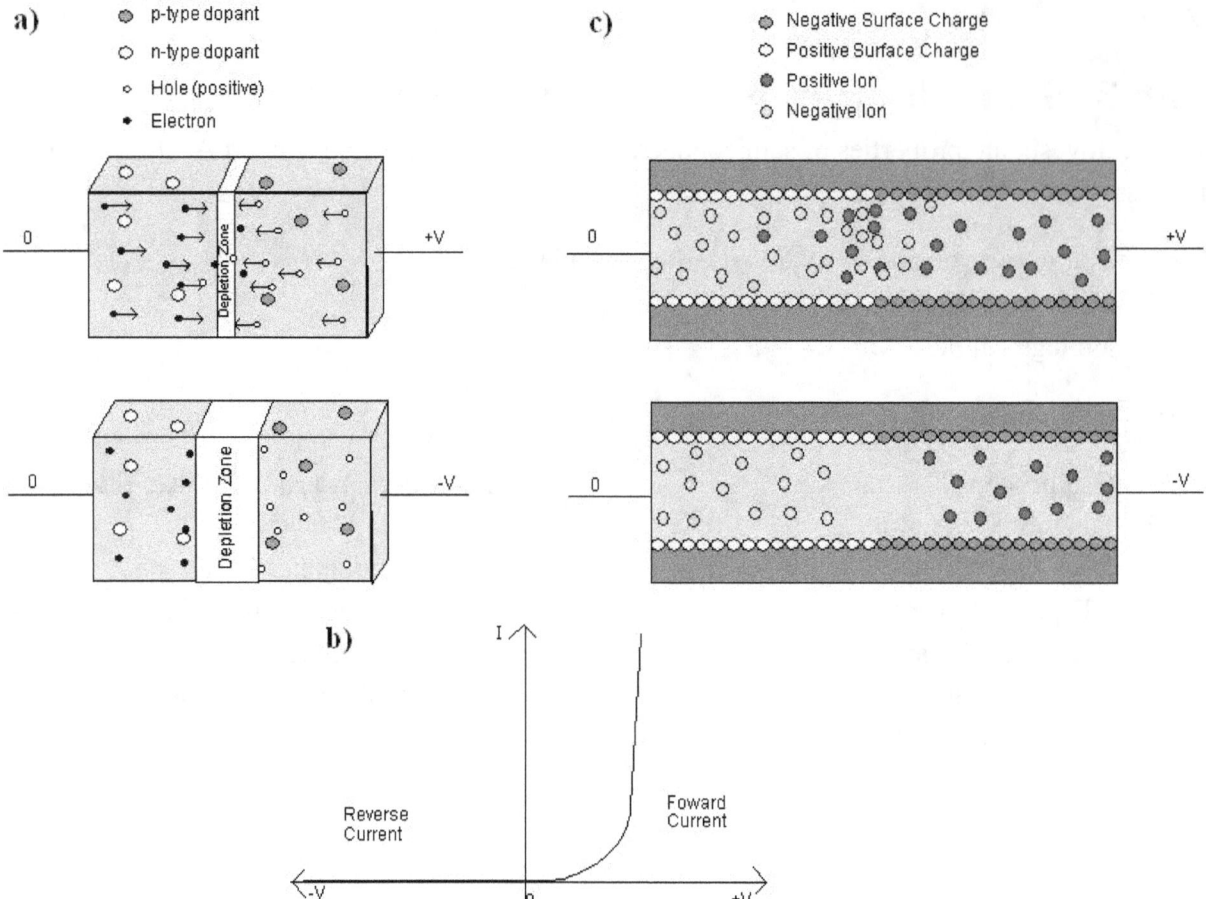

Figure 3. 1: A forward-biased pn junction (a) showing diffusion of charge (current) into the depletion zone. Note the reverse-biased pn junction showing no diffusion. Panel (b) shows the corresponding I-V curve. Note the strong non-linearity of the response at Vd. Panel (c) shows a cutaway view within an actual diode showing the surface charge surrounding the conduction channel.

Two rules for diodes:

1) When forward biased, if $V > V_d$ then the current that flows is whatever is required for $V = V_d$.
2) When forward biased with $V < V_d$ or reversed biased, the voltage drop across the diode is whatever voltage is applied.

Limitations to Diodes

There is a limit to how much current a diode can pass when forward biased. Since $P = IV_d$, it is not surprising that as the power increases, it will reach a limit where the diode will overheat and breakdown. When it is reversed-biased, there exists a breakdown voltage (V_{br}) where the small number of charge carriers can acquire sufficient momentum to knock loose both electrons and holes creating an avalanche of (backward) current. In most circuits, it is important to insure that this breakdown voltage is not reached. However, this phenomenon is exploited in avalanche photodiodes to produce a current pulse when photons of light are absorbed at the pn-junction interface and triggering avalanche breakdown.

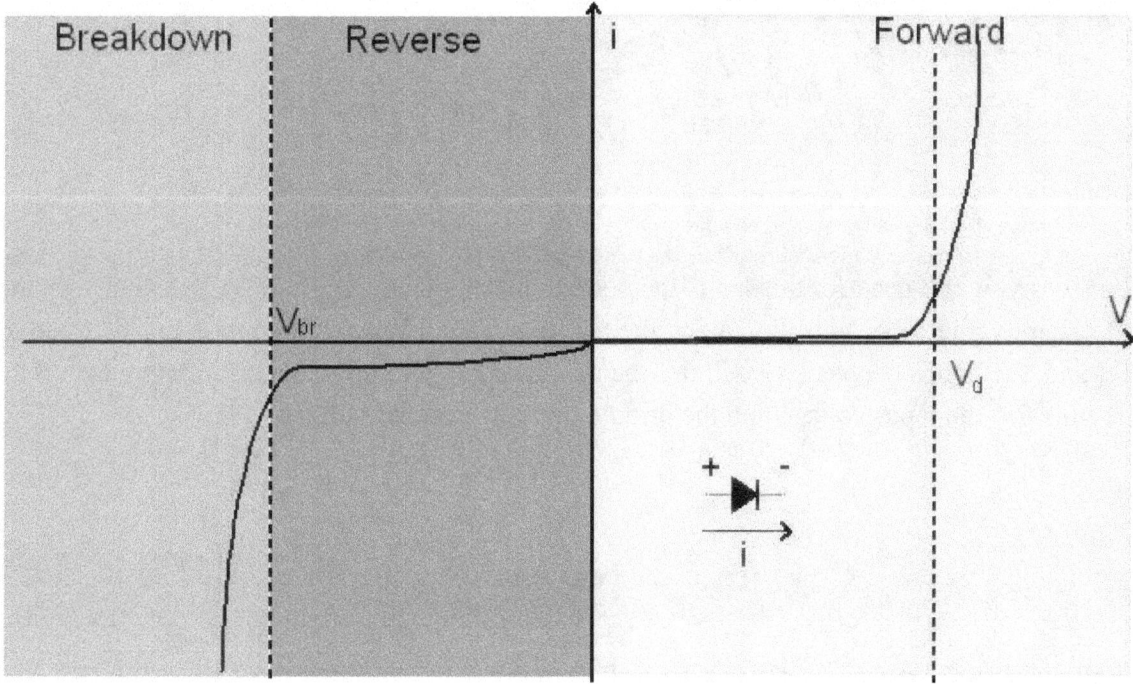

Figure 3. 2: The I-V curve of a pn-junction diode. Note that a minimum positive voltage (V_d) is required for current to flow. Note further that for reversed biasing there is still a small amount of current but this is nearly constant with voltage until the breakdown voltage (V_{br}) is reached; at which point current flows (backward). In a real diode V_{br} is considerably more negative than V_d is positive. The diode circuit symbol is also shown with the arrow pointing in the direction of traditional current flow.

Zener Diodes

Zener diodes also make use of the avalanche breakdown to provide a source of reference voltage. That is, if a Zener diode is placed in a circuit it can allow current to flow precisely when V_{br} is reached and, thereby, serve as a voltage standard within the circuit. We will see some examples below.

Equipment Notes

Potentiometers

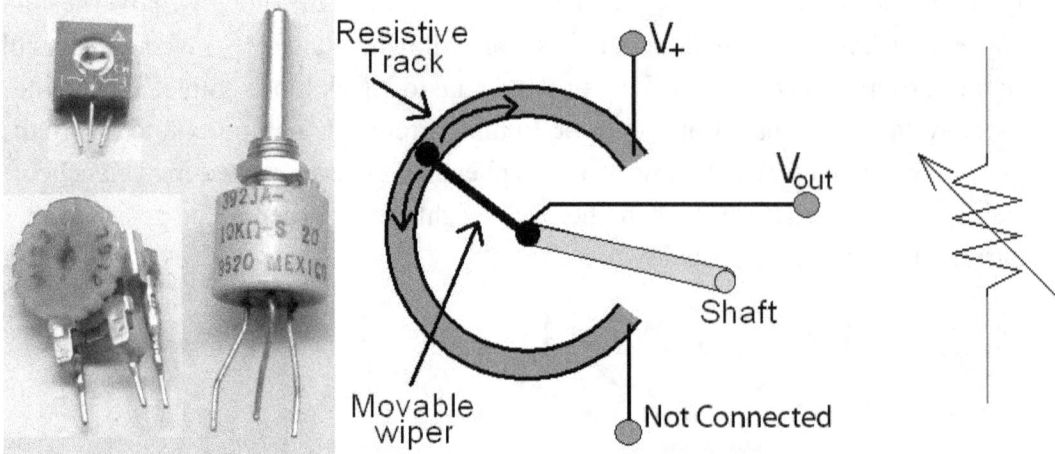

Figure 3. 3: (left) Three examples of potentiometers. (center) The internal structure of a single turn potentiometer. Either of the outer pins can be connected to the input voltage when using the potentiometer as a variable resistor. The middle pin is, then, Vout. (right) The circuit symbol for the variable resistor.

One pieces you will need for this lab is the potentiometer shown in Figure 3. 3. As you turn the shaft, the movable wiper will slide along the resistive track, changing the resistance between the wiper and V_+. In this chapter, we will use the potentiometer as variable resistor using one of the outer pins for the input voltage and the middle pin as the output voltage.

Diodes

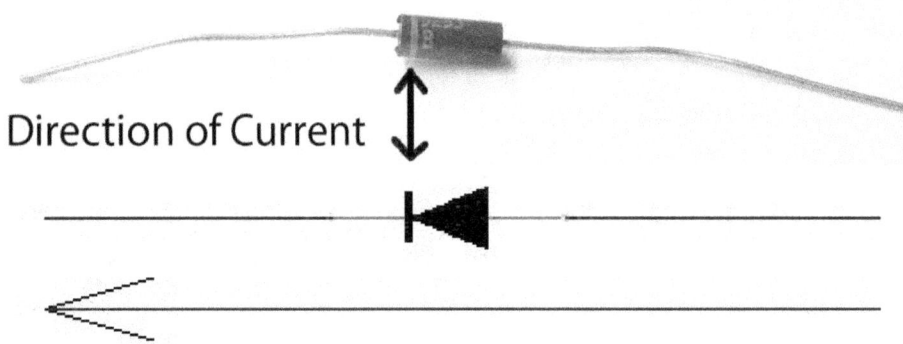

Figure 3. 4: The diode and how to determine direction of current flow.

For the diode, direction of current flow matters. The direction of the symbol shows how the diode should be connected in the circuit as well as the direction of current flow is shown in Figure 3. 4.

Procedure

Part I: I-V Curve of Diodes

Let's begin by looking at the behavior of diodes in simple circuits. You will build a voltage divider like the one shown in Lab 1 but include a 1N4001 diode in parallel with R_2 (Figure 3. 5) and a small load resistor (R_L) to limit the current into the circuit. If we connect the +12V power supply (not the FG) to the breadboard so that V_{in} is 12 volts, we expect an output voltage of:

$$V_{out} = V_{in}\left(\frac{R_2}{(R_L + R_1) + R_2}\right) \qquad (3.1)$$

Now if we make R_1 a potentiometer (used as a variable resistor), we can see what happens as it is varied. If the pot is turned up so that R_1 is greater than the value you found in Prelab Question (1) (a), then the output voltage across the diode will be less than its V_d. For an ideal diode, no current will flow through the diode causing the voltage across R_2 to be equal to the pure voltage divider Equation (3. 1). However, if we slowly decrease R_1 then the voltage across the diode will increase until $V_{out} > V_d$. At this point current will flow and begin bypass R_2 so that the output voltage across the load will always be $V = V_d$.

Real diodes are much less clearly defined, however, and require more investigation to see exactly how they respond to a slowly increasing input voltage.

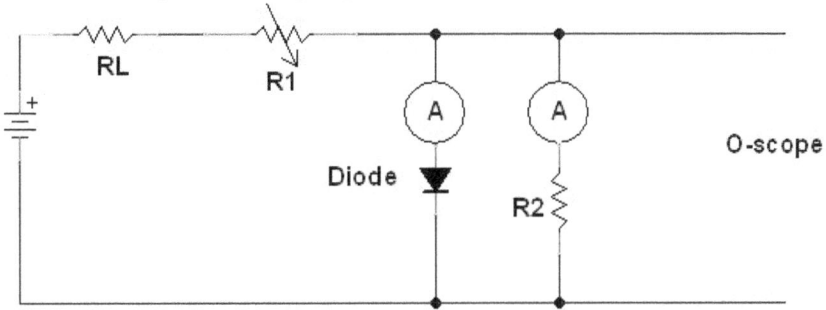

Figure 3. 5: Voltage Divider with Diode across the output.

- Construct this circuit in Figure 3. 5 with $RL \approx 100\Omega$, ½ watt, $R1 \approx 10k\Omega$ potentiometer, and $R2 \approx 200\Omega$ 1/4 watt.
- Use two DMMs as ammeters placed where shown in the circuit, both set to *mA* range. Use alligator clips to connect the probes to wire leads from the breadboard, and remember that current must be measured by placing the ammeter in series with the circuit.
- Monitor the voltage across R_2 with the o-scope as the resistance in the pot is altered. At the same time, measure the current through R_2 and the diode with the DMMs. In the interest of time, make measurements as follows: 5 measurements from 0V to 0.5V, 5 measurements from 0.5V to 0.7V, and 5 measurements from 0.7V to 1.4V.
- From the data, find the voltage at which the current through the diode surpasses the current through the resistor and record it. This is the turn-on voltage, V_d.

- ➢ After lab, make an I-V curve like shown in Figure 3. 2, but include both the resistor and diode current on the one graph. Make both a linear-linear graph and a log-linear graph to better see the huge range of current.
- ➢ Examine the nearly linear sections of each curve (diode and resistor) before and after the diode activates (the first and last 5 measurements taken for each). Make a linear fit for these four sections. Discuss what the slopes of the fits mean, as well as the meaning of the change from before the activation to after.
- ➢ If you were designing a safety circuit to protect a sensitive piece of equipment (R_2) that had a max voltage rating of 0.6V, would you design this circuit? Why or Why not? Look up germanium diodes. Would germanium diodes be a better or worse choice?

Part II: I-V Curve of Zener Diodes

If you want to limit the voltage across sensitive components, but need a much higher voltage, a Zener diode is the way to go. Zener diodes are made with a variety of breakdown voltages that, when reached, allows current to flow backwards across the diode. Unlike a regular diode, this will not destroy it, and the voltage across the diode will remain fixed at the breakdown voltage (for an ideal Zener diode anyway).

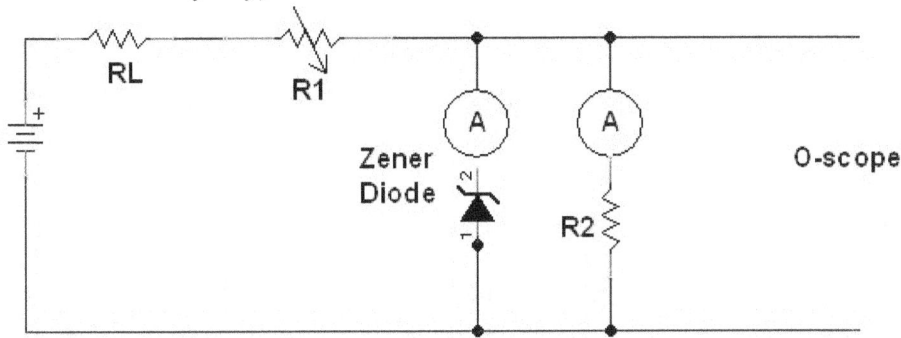

Figure 3. 6: A Zener diode is used to limit the voltage across R_2 to V_d.

We will investigate this use by replacing the 1N4001 diode with the 1N4733 Zener diode. Be sure to reverse its direction as shown in Figure 3. 6. In this case, if R_1 is lowered to the value you found in Prelab Question (2), then the voltage across the diode will exceed the breakdown voltage allowing most of the current to flow through the Zener diode. This means that the voltage across R_2 can never exceed the Zener breakdown voltage (for ideal diodes), protecting the resistor from having too much voltage, or voltage spikes.

- Since this Zener diode has a high V_d, use the 12V input to the circuit. Change R2 to a 1kΩ resistor and monitor the voltage across R2 and the currents through R_2 and the diode as you did in the previous section.
- Investigate the response of the Zener diode as you did in the previous section. Take 15 measurements between 3V and 6V. Locate the breakdown voltage for the diode as you did before.

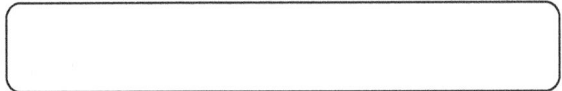

- Graph the data the same way you did for the 1N4001 diode, and examine the slopes before and after the diode activates for both the resistor and the diode. Does the breakdown voltage behave the same as the turn-on voltage for the other diode, or is there some difference?

Part III: Rectifying an AC Voltage: Half-Wave Rectifier

A common problem in circuit design is the creation of a direct current (DC) power source. Most modern electronics make use of semiconducting devices and these electronics need a stable source of DC power. Most power, however, is provided in alternating current (AC) form due to its greater efficiency at long transmission ranges. One common way of converting AC power to DC power makes use of diodes.

Diodes allow for a process known as rectification, where AC can be converted to DC. Rectifiers are in a variety of devices, most notably in the AC to DC conversion boxes on the power cord of a laptop computer or HDTV. In this setup, generally, a transformer provides an efficient means for stepping down the transmitted line voltage to a level used by household electronics. The voltage can then be "rectified" using diodes to become DC.

The simplest form of rectification takes advantage of the nonlinearity of the diode to allow current in one direction and thus only positive voltage remains in the rest of the circuit. This circuit is called the half-wave rectifier because the negative half of the waveform is removed from the output and dissipated through resistive heating of the diode. The result is an irregular waveform (Figure 3. 7). The pulses mean that it has large "ripple", a measure of the variation in the DC voltage. Ripple is usually expressed as a percent of the average level.

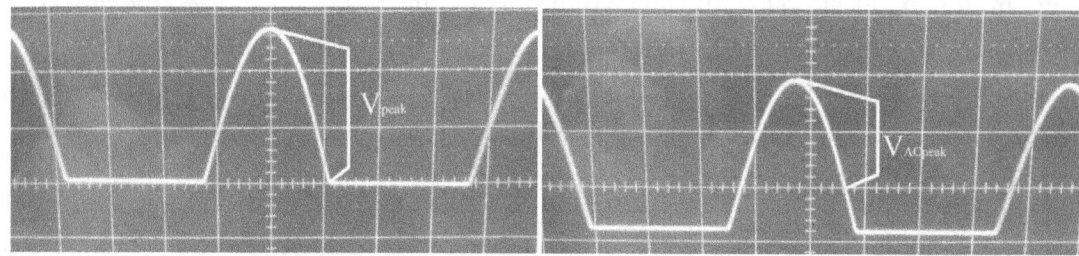

Figure 3. 7: Half-wave rectification of an AC signal. (left) panel shows a rectified sine wave with the o-scope set on DC. (right) the same wave but with the o-scope set to AC).

The output voltage of the half-wave rectifier does not produce the same voltage of the input as measured by the peak voltage V_{peak}.

The relationship between the peak voltage and the rms voltage is no longer related by a factor of $\sqrt{2}$ as seen in Lab 1. Here is the equation to find the average voltage.

$$V(t) = V_{peak} * \sin(t) \quad (3.2)$$

There are 2π radians in a full circle, and the half-wave rectifier will use only half of that.

$$V_{avg} = \frac{V_{peak}}{2\pi} \int_0^\pi \sin(t) \quad (3.3)$$

And after solving and simplifying yields,

$$V_{avg} = \frac{V_{peak}}{\pi} \quad (3.4)$$

The RMS of the Voltage is also different from that of the AC signal from the FG. We find the RMS voltage by integrating over the waveform, where $V_{AC\,peak}$ is the difference from the

$$V_{rms} = \sqrt{\frac{1}{2\pi} \int_0^\pi \left[V_{AC\,peak} \sin(\omega t)\right]^2 d(\omega t)} \quad (3.5)$$

And again after some simplification,

$$V_{rms} = \frac{V_{AC\,peak}}{2} \quad (3.6)$$

Because the output is DC but with an AC component, we define the voltage ripple (V_{ripple}) as the peak-to-peak voltage of the AC component of the signal. See Figure 3.12 for example of approximately 20% ripple.

$$V_{ripple} = \frac{I_{Load}}{fC} \quad (3.7)$$

Where f is the frequency from the FG, C is the capacitance of the capacitor used for smoothing, and I_{load} is the AC current through the resistor while the capacitor is removed.

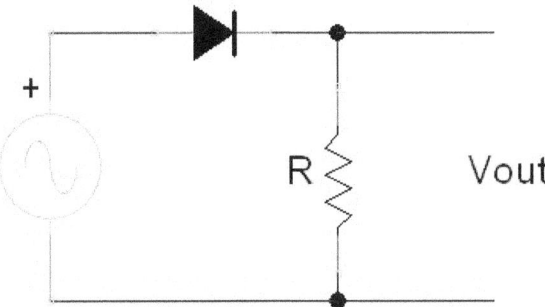

Figure 3.8: The half wave rectifier circuit. The order of the diode and resistor are important. If reversed, the signal will have a negative component.

- We will use the 1N4001 diode and a resistor of at least 1000 ohms. Set up the circuit shown in with the o-scope connected across the resistor (be sure the positive lead is on the positive side of the resistor). Also, check that the o-scope channel reading across the resistor is set to DC. Set the FG to medium amplitude and feed the FG output into the other channel of the o-scope.
- Compare the two waveforms. Make a quick sketch (or take a picture) of this and record the voltage of the peaks of each waveform. Be careful to measure the peak voltage of the rectified signal from the flat portion to the peak. There should be a difference between this value and the peak-to-zero voltage on the input signal; what is the amount?

Introduction to Semiconductors: Diodes   Introduction to Basic Circuits and the Arduino

> ➢ Lower the amplitude of the FG by about half and measure the difference between the peak voltages. Is it the same as before? What you just measured is the voltage drop across the diode (V_d). Compare this value to the value calculated in Part I of this lab.

> ➢ Your DMM calculates the V_{rms} by first treating the rectified DC signal as an AC signal, and adjusting the signal so that the waveform oscillates above and below the zero point. To see how the DMM views the wave, switch the o-scope to AC, notice how part of the rectified signal drops below the zero line on the screen. Measure the V_{ACpeak} (Figure 3. 7) and record the value below. Measure the RMS voltage of the rectified signal using the DMM. Does the RMS voltage of the rectified signal agree with Equation (3. 6)?

> ➢ Now switch your DMM to measure DC and record the average voltage across the resistor. Does your answer agree with Equation (3. 4)?

Part IV: Rectifying an AC Voltage: Full-Wave Rectifier

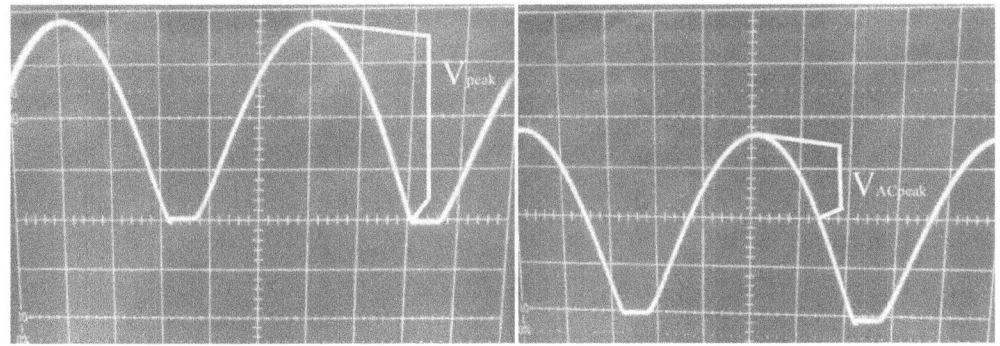

Figure 3. 9: (left) Full wave rectified signal with o-scope set to DC. (right) Same signal with o-scope set to AC.

By adding additional diodes, we can flip the negative portion of the AC signal to become positive. This rectifier is known as a full-wave rectifier which, though slightly more complicated to build, converts the entire waveform into DC (Figure 3. 9). This has the advantage of being more efficient since none of the current from the AC signal is thrown away. However, it still has a significant amount of ripple.

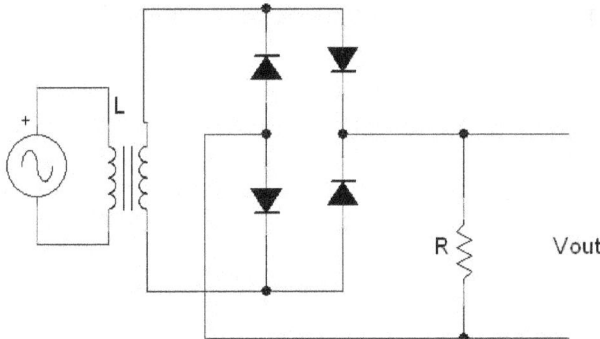

Figure 3. 10: A full-wave rectifier circuit with $R > 1k\Omega$. Note that the current through the resistor will be unidirectional.

> You will now make a Full Wave Bridge Rectifier. This circuit will require four 1N4001 diodes arranged as shown in Figure 3. 10 as well as an isolation transformer that is center tapped to isolate the FG ground from the o-scope ground.

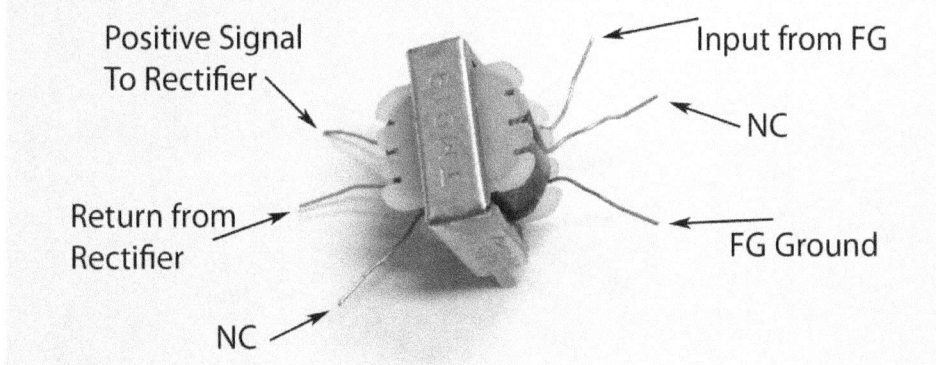

Figure 3. 11: An isolation transformer and the leads to use.

- ➤ Before inserting the transformer, determine the turn ratio using the technique in Lab 2 Part III. This time, however, measure across the center tap as shown in Figure 3. 11.

  [ ]

- ➤ When attaching the transformer in the rectifying circuit, feed the FG input and ground into the corners on one side while the output will come from the center wire (pos/neg) and one corner wire (return) (Figure 3. 11).

- ➤ Now that the ground from the FG is isolated from the o-scope, attach the o-scope probe across the load resistor (complete with the ground clip). If your rectified signal is a flat line, check your calculation to be sure that the output voltage from the transformer is high enough to activate all the diodes.

- ➤ Adjust the amplitude of the wave up and down observing the effect on the rectified signal. What happens when the amplitude is very low?

  [ ]

- ➤ Does the peak voltage match what you expect given the transformer you chose? Remember to account for the voltage drop across the diodes as you found in Prelab Question (3).

  [ ]

- ➤ Take a picture of the waveform.
- ➤ Measure the average (DC) voltage and the V_{rms}. Do these values match the equations you found in Prelab Question (4).

  [ ]

- ➤ In your rectified signal, does the base of the peaks meet each other, or is there a flat line at zero volts between each peak? In your report, explain why this flat line exists. Also explain what happens to these flat lines when you increase the amplitude and why.

Part V: Buffered Full-wave Rectifier

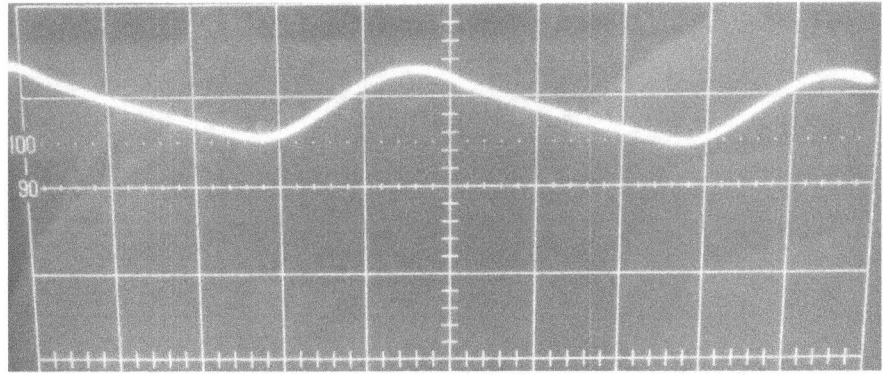

Figure 3. 12: The rectified signal after a filtering capacitor has been applied.

If we add a capacitor to the output of a full-wave rectifier, we have added a high-pass filter (Lab2). This new circuit is called a buffered full-wave rectifier. The high-pass filter has the effect of smoothing out the pulses (ripple) from the rectified wave as seen in Figure 3. 12. Depending on your specific needs multiple circuits can be chained together to more fully buffer and smooth the output signal.

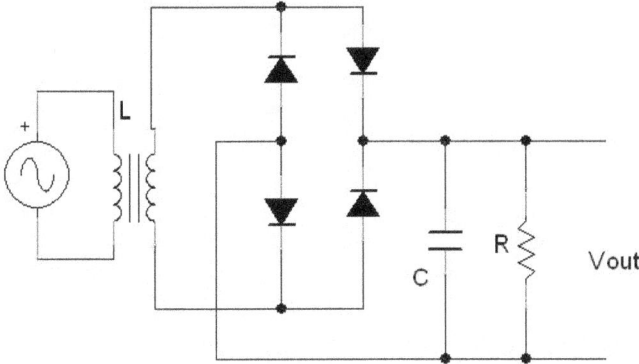

Figure 3. 13: buffered full-wave rectifier. The left panel shows the circuit diagram and the right panel shows the output.

> First, set the FG to 1000 Hz and measure the AC current through the resistor without the capacitor in place. Then, using Equation (3. 7), calculate an appropriate value for the capacitor that will bring the ripple down to 10% of the peak voltage. Keep in mind that for a full-wave rectifier, the frequency, f, in the equation is doubled!

> Connect the o-scope across the resistor and record the new waveform and Average DC voltage with the DMM. How does the DMM voltage compare to the DC voltage read on the o-scope? Name two ways to decrease the V_{ripple}.

Note that the simple "wall-warts" that you plug into AC power to provide DC power (Figure 3. 14) are not much more complicated than the circuit we built. The only significant additions are that of a step-down transformer that takes the place of the signal generator and a voltage regulator to further smooth the signal.

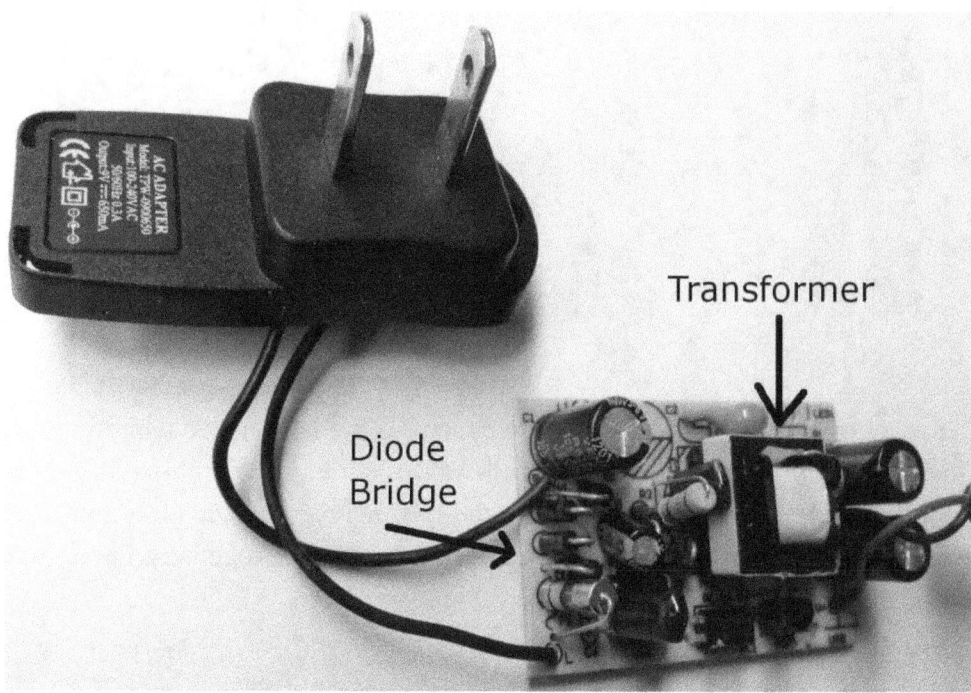

Figure 3. 14: The internal circuitry of an AC-to-DC conversion box (wall wart). The simplest of these are nothing more than a buffered full-wave rectifier.

Part VI: Voltage Regulator

The final addition that is often made to an AC adapter like the one you just built, is the addition of a voltage regulator. A voltage regulator is an integrated circuit that is specifically designed to produce a DC signal with an extremely low ripple. Also, with the LM317 adjustable voltage regulator, it is easy to produce the exact voltage you desire with high precision.

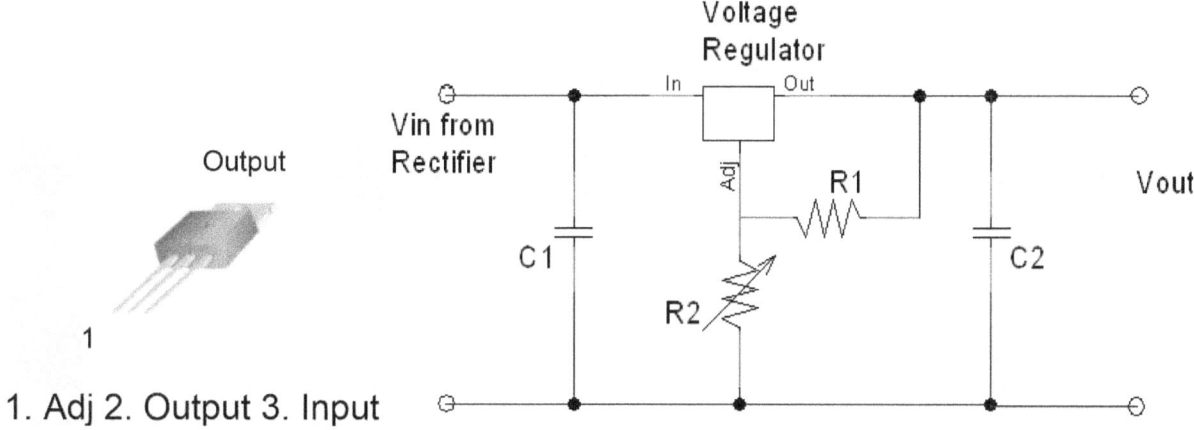

Figure 3. 15: (left) the LM317 and its pinout. (right) The circuit to be built around the LM317 to produce the correct output.

- Adjust the voltage on your FG so that the output voltage from your bridge rectifier is between 7 and 12 Volts. We do this because a good rule of thumb is that the input voltage into a regulator should be at least 2 Volts greater than the voltage you want out of the regulator.
- Build the circuit in Figure 3. 15**Error! Reference source not found.** using the output from your bridge rectifier as the input. C1=0.1µf, C2=1.0µf, R1=240Ω, and R2=1000Ω potentiometer.
- Keep in mind that the LM317's pinout is arranged in a different order than in the circuit diagram. Also, be sure to set the potentiometer (R2) to a middle setting before connecting the power.
- Adjust R2 while reading the voltage out with the DMM until you get a value just below 5 Volt. This is a good value to have because the Arduino runs on 5 Volts and anything more is a danger to the board.
- Switch to measuring the output with the o-scope and measure the ripple, or determine the maximum ripple value given the limitations of the o-scope.
- When finished, raise and lower the voltage from the FG and watch what the output voltage from the regulator does. Does the regulator deliver a constant voltage independent of the input voltage?

Part VII: Summary

In this lab, you have examined the properties of diodes. You have also investigated some of the uses of diodes in simple circuits. In particular, you have used diodes to make a simple DC power supply similar to those used in electronic devices.

Part VII: Further Applications

i. **Signal Diodes:** Diodes can be connected across a signal line that acts as a kind of release valve to allow transient signals to be filtered out.

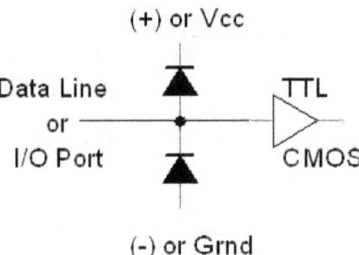

Figure 3. 16: If a voltage spike travels down the data line, it will be diverted through the diodes to either the ground or Vcc before it can reach the end.

ii. **Signal Diodes in Series:** If used in a series, you can create a simple voltage regulating circuit where the voltage after each diode is dropped from the previous diode by the voltage drop across the diode. Since these are fairly constant, the voltage output at each step is easy to predict.

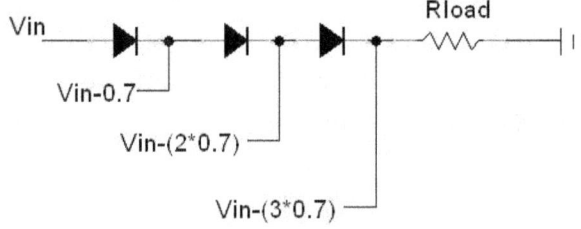

Figure 3. 17: Multiple signal diodes can form a simple voltage regulator.

iii. Pulse Width Modulated (PWM) LED: If you want to increase the brightness of an LED, but cannot supply more power to the LED without destroying it, a PWM signal with peak voltage higher than the LED's voltage rating will produce more light. The LED is not destroyed because the gaps between the peak voltage in the signal keep the average voltage within a safe range. If the frequency is high enough, your eye will not be able to tell that the LED is blinking, and it will appear brighter.

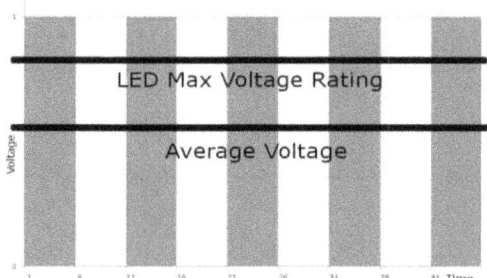

Figure 3. 18: The Pulse Width Modulated signal can increase the apparent brightness of the LED without destroying it.

iv. Opto-coupling: If you need to electrically isolate one circuit from another, you can use an LED and a Photodiode as a switch. The LED will light when there is sufficient power while the Photodiode will allow current to flow when it senses light. This can also be done with a Phototransistor (Lab 4).

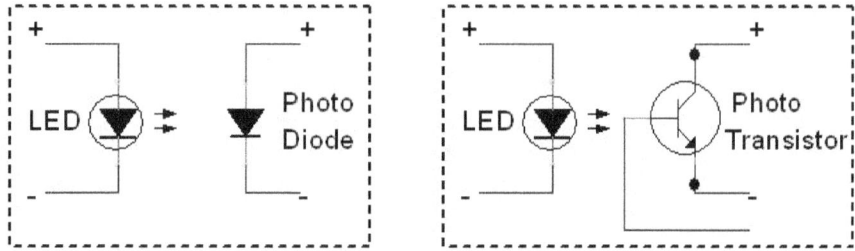

Figure 3. 19: Opto-couplers using a Photodiode or a Phototransistor.

v. High Precision Voltage References: Some types of zener diodes are designed to deliver a specific voltage with high precision and accuracy. These diodes can be placed in parallel to feed a reference voltage into any voltage measurement device (like an Arduino or another ADC).

vi. Battery Backup of Electronic Devices: If an electronic device is attached to a DC power supply and battery through diodes (which keep each one isolated) we can use the diodes to switch-in the battery backup in the event of a power failure. When the AC/DC power fails the diode between the device and the battery is no longer reverse-biased and so the battery will discharge and provide temporary power to your electronic device. Note the second diode that prevents battery current from going upstream into the DC power supply. That is, you can keep playing WOW or COD long enough to save progress and shutdown your computer. When the AC comes back up the diode next to the battery is now reverse-biased and the battery will stop discharging.

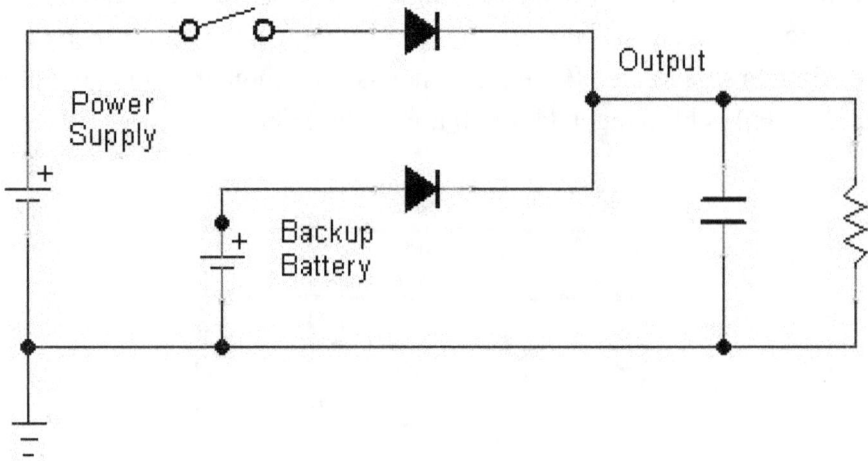

Figure 3. 20: A cartoon diagram of a battery backup circuit.

For more information, check out this site: www.electronics-tutorials.ws

Part VIII: Prelab Questions

(1) In Part I (1N4001 diode), the circuit consists of a voltage divider that delivers a variable voltage to the diode.
    a. Calculate the of value of R_1 necessary to bring $V_{out} = V_d$, where $V_d = 0.6$ volts for the 1N4001 diode.
    b. Calculate the min and max voltage emanating from the voltage divider when the potentiometer is adjusted.
    c. For an ideal diode, what would be the max voltage across the resistor, R_2.

(2) In Part II, calculate the value of R1 necessary to bring Vout = V_b for the Zener diode, where $V_b = 5.1V$

(3) In Part IV, you will be building a bridge full wave rectifier.
    a. Draw two circuit diagrams of the bridge full wave rectifier and trace the two paths the AC signal follows through the circuit (one on each diagram) to show that the current through the resistor is always in the same direction.
    b. Considering the paths you found in (a), what is the total voltage drop due to the diodes once the current reaches the resistor.

(4) In Part IV, you will need to measure V_{avg} and V_{rms}. Starting with equations (3.3) and (3.5), decide on the appropriate integration limits and derive the equations for these two quantities for a full wave rectifier.

Lab 4 - Introduction to Transistors

Goals: Investigate properties of the bipolar junction NPN transistor through the three most common configurations and use it to construct a common emitter amplifier. You will study:

1. Common Base, Common Collector, Common Emitter
2. How to Determine the Operational Point
3. An AC Amplifier

List of Equipment and Parts

1. Oscilloscope
2. Function Generator
3. 2 Digital Multimeters
4. DC Power Supply
5. Solderless Breadboard
6. Jumper Wires
7. 3 Capacitors (2000pf, 0.47μf, 470μf)
8. 11 Resistors (two 680Ω, 1kΩ, 5kΩ, 10kΩ, 100kΩ, 430kΩ, three 500kΩ, 620kΩ)
9. 2 Potentiometers (10kΩ, 500kΩ)
10. 1 Transistor (2N2222)
11. 1 LED
12. 3 Banana wires
13. 3 Alligator clips

Introduction and Background

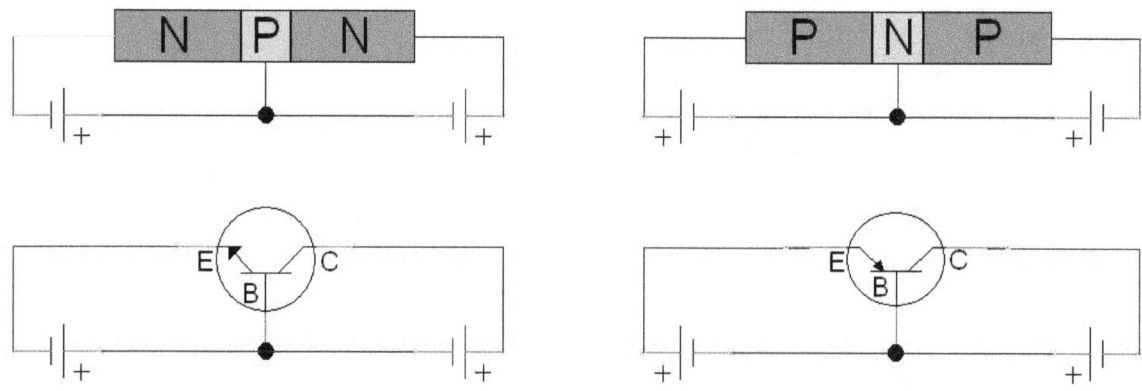

Figure 4. 1: (left) the npn transistor, and (right) the pnp transistor. Note that VBE reverses.

We begin with a summary of the bipolar junction transistor. The device is basically a sandwich of three different semiconductor layers. There are two types: a *negative-positive-negative* (npn) sandwich, and a *positive-negative-positive* (pnp) sandwich. Figure 1 shows the structure of the bipolar junction transistor. Both type have three leads, *an emitter, a base, and a collector* but in both types the center post is the base.

Consider the npn transistor. When the device is forward biased ($V_{BC} = V_B - V_C < 0$). V_{BC} is usually about 0.65 V. The connection between B and C looks like a diode so no current flows from B to C. As a result, $V_{CE} = V_C - V_E > 0$. Thus we also know that $I_E = I_B + I_C$. The transistor acts like a valve for current and with the base voltage acting as the valve. A small Base current controls a much larger current flowing into the collector.

This behavior leads to many applications including simple high-speed switching to amplification. In amplification applications, the transistor receives and analog signal, such as an audio signal, through its collector and a separate signal is applied to its base. The output at the emitter will be identical to that at the collector but amplified by an amount proportional to the base (control) voltage.

Specifically:
$$I_c = \beta I_B \tag{4.1}$$

with β being 100 – 300. Similarly we define α so,
$$I_c = \alpha I_E \tag{4.2}$$

Combining,
$$I_B = (1 - \alpha) I_E \tag{4.3}$$

So we have:
$$\beta = \frac{I_C}{I_B} = \frac{\alpha I_E}{I_E - \alpha I_E} \tag{4.4}$$

and thus,
$$\beta = \frac{\alpha}{1 - \alpha} \text{ and } \alpha = \frac{\beta}{1 + \beta} \tag{4.5}$$

Typically, $\beta \gg 1$.

There are several figures that can illustrate the behavior of the transistor, but Figure 4. 2 shows the collector current (I_C) as a function of voltage between the collector and the emitter (V_{CE}). There is a different I-V curve for each value of the base current (I_B). The rapid rise shows the transistor turning on as V_{CE} is increased, after which the current (I_C) is approximately constant. The central region, where I_C is roughly proportional to I_B is known as the *linear active region*. The rapid rise on left side is called the **saturation region**. The dissipated power ($P = I_C \times V_{CE}$) must be less than the maximum power rating for the transistor and this is plotted as the dotted curve.

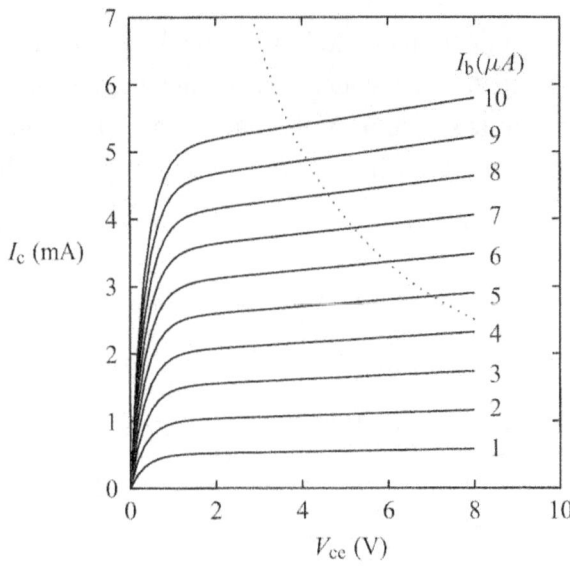

Figure 4. 2: An I-V curve for an npn transistor

Equipment Notes

Transistors

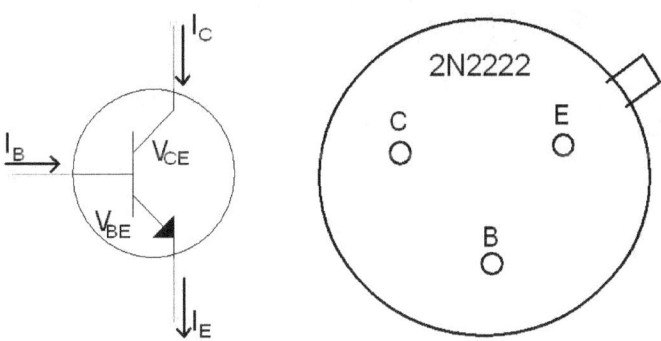

Figure 4. 3: Package for the 2N2222 transistor. The component diagram is shown on the left and the pins identified on the right hand panel. Transistor is viewed from the bottom.

Transistors use a signal with a small current at their base to control a larger current (and possibly voltage) through the collector and emitter. We will be using the 2N2222A, a NPN transistor. Study Figure 4. 3 and notice the relative positions of the collector, emitter, and base. Keep in mind that this figure is a transistor as viewed from the bottom of the posts, not the top of the can. Failure to place the posts correctly on your breadboard can destroy the transistor!

To keep your circuit organized on the breadboard, use the top and bottom horizontal rails as power and ground and build your circuit with perpendicular jump wires wherever possible. Finally, since placement of the transistor is critical, place the transistor in the breadboard first as shown in Figure 4. 4, and use multiple colors to indicate each pin. Once the transistor is in place, build the circuit out from the ends of the jumper wires so that the transistor orientation never changes.

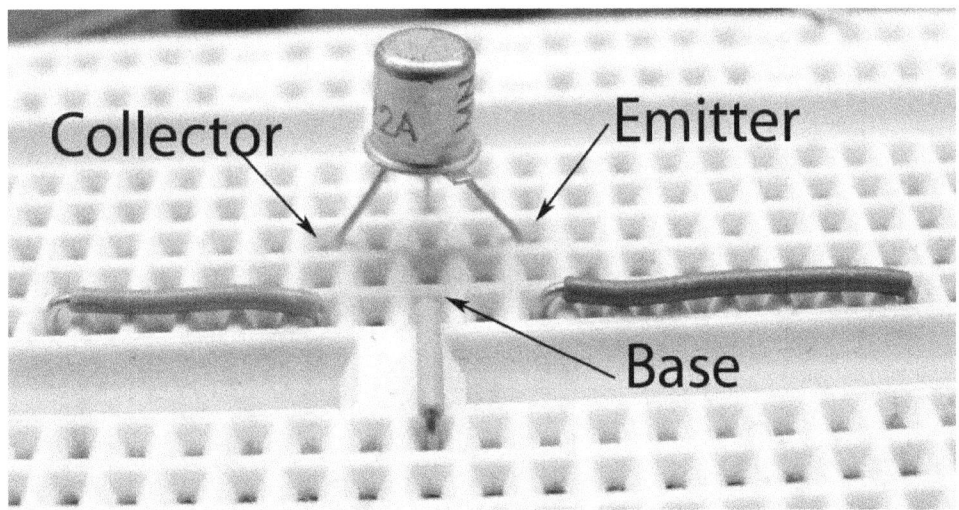

Figure 4. 4: The proper way to place the transistor in the breadboard.

Potentiometers

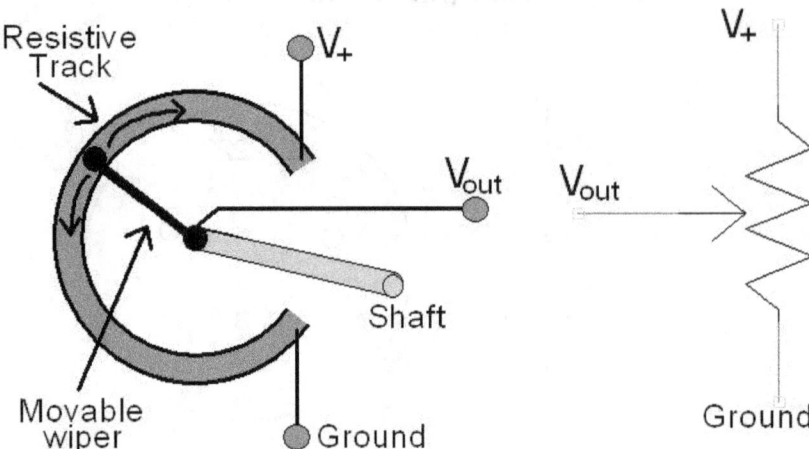

Figure 4. 5: (left) The internal structure of a single turn potentiometer. The outer pins are input voltage and ground. The middle pin is the Vout of the voltage divider created by the resistive track. (right) The circuit symbol for the potentiometer.

Until now, we have only used one side of the potentiometer as a variable resistor. If both sides are used, however, the potentiometer becomes a voltage divider complete in one little package. Looking down at the pins, we can see in Figure 4. 5, if we connect a voltage across the two outer most pins, then the center pin is the output voltage from the voltage divider. We can now use the potentiometer to vary an input signal in a smooth fashion.

Part I: Common Base, Collector, Emitter

Transistor circuits come in three basic varieties; Common Base, Common Collector, and Common Emitter. We will explore these three setups starting with the common base.

Part Ia: Common Base

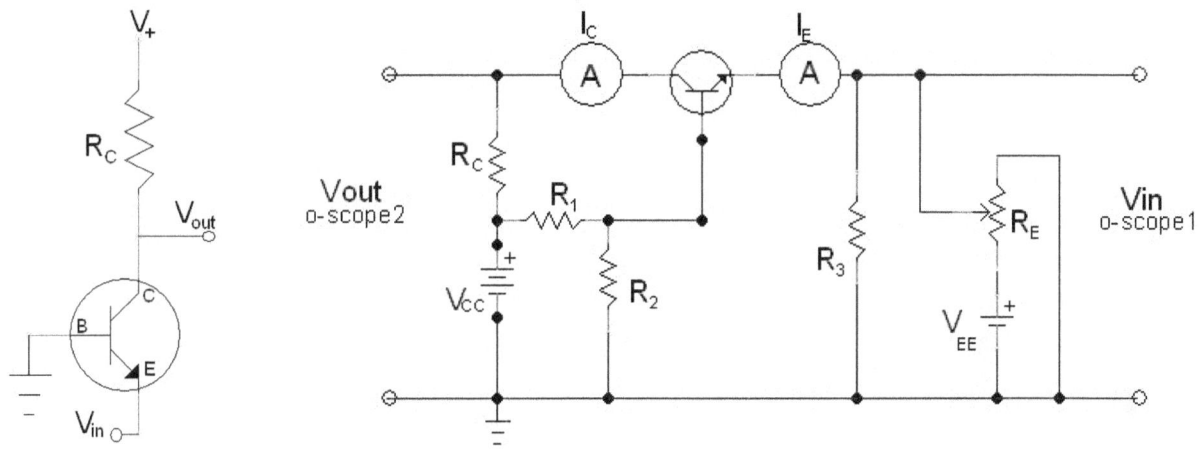

Figure 4. 6: The Common Base Amplifier. (left) The simplest circuit diagram with no consideration of biasing. (right) A functional circuit diagram. Note that in an AC amplifier, the FG would replace V_{EE}.

The common base is so called because both the input signal and the output signal share a common point at the base (Figure 4. 6). Notice that, for both the DC and AC amplifier, the base is forward biased by the V_{BB}. The input signal, however, is measured across the emitter side, V_E, while the output signal is measured across the collector side, V_C.

- Setup the circuit shown in Figure 4. 6 where R_C = 5kΩ, R_3 = 500kΩ R_1 = 100kΩ, R_2 = 10kΩ, and R_E is a 10kΩ potentiometer (used as a voltage divider). Connect V_{CC} to +12V and V_{EE} to +3.3V.
- Connect your two DMMs to measure current I_E and I_C. Connect the o-scope probes to read V_C and V_E as shown.
- There is a limited range over which the transistor behaves linearly. Examine the output voltage on the o-scope. As R_E is adjusted, the boundaries of your measurements will be given by where the output voltage reaches a maximum and a minimum and no longer moves (Figure 4. 7). Record the max and min for I_E, I_C, V_E, and V_C.

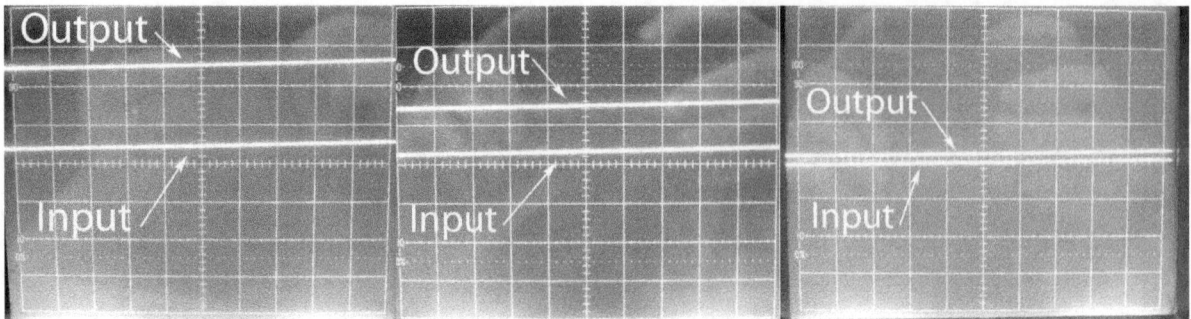

Figure 4. 7: The progression of the output and input voltages from the common base as the potentiometer R_E is adjusted. Maximum (left), Intermediate (middle), Minimum (right). Note that the Input is on a 1Volt scale while the Output is on a 5V scale.

➤ As you adjust R_E, record I_E and I_C with the DMMs, and record V_C and V_E with the o-scope. Stay within the bounds you found in the step above. Remember that the relation between currents and the relation between voltages are both linear in the operational range of the transistor. You need only measure enough data points to get a good measurement of the slope of line. If your data turns non-linear, then you have exceeded the bounds of this experiment.

➤ Make a graph of I_E vs. I_C to find the value of α (Equation (4. 2)). Based on your answer, is the common base amplifier a good current amplifier?

➤ Make a graph of V_C vs. V_E. Based on your data, is the common base amplifier a good voltage amplifier?

Part Ib: Common Collector

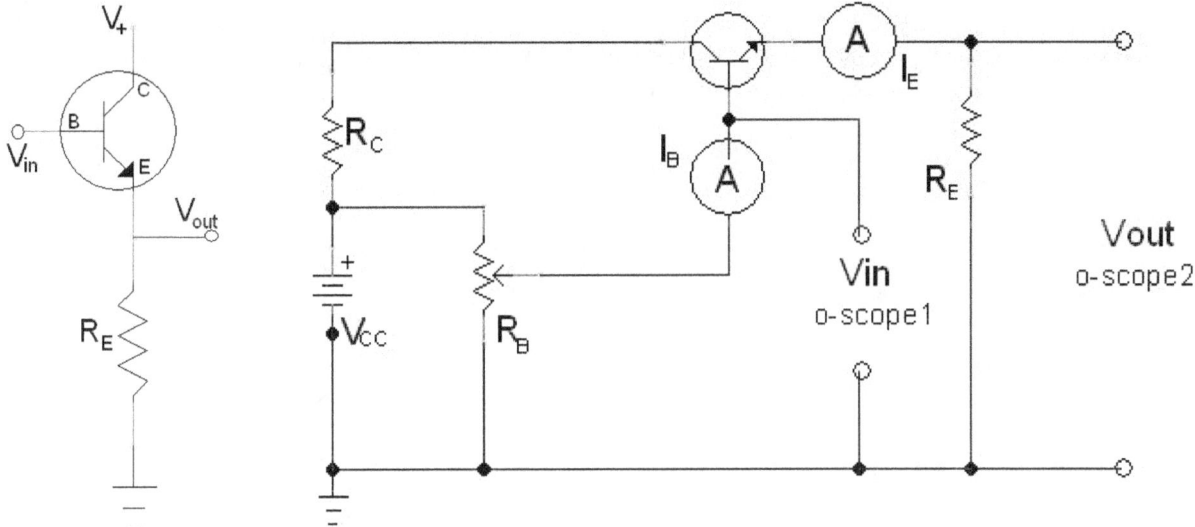

Figure 4. 8: The Common Collector Amplifier. (left) Simplest circuit design with no consideration of biasing. (right) A functional circuit diagram. Note that in an AC amplifier, the FG would be placed between the ammeter and ground.

The common collector is so called because both the input signal and output signal share a common point at the collector side. An input signal feeds a small current into the base of the transistor, which acts like a control valve for the current from V_{CC}. Input is measured from the base, while output is measured from the emitter side of the transistor.

➢ Setup the circuit shown in Figure 4. 8 where R_E = 1kΩ, R_C = 5kΩ, and V_{CC} is +12V. R_B is a 500kΩ potentiometer used as a voltage divider. Refer to the Equipment section on how to use it if you are confused.
➢ Connect the DMMs to measure the currents I_E and I_B as well as the o-scope to measure V_E and V_B.
➢ Adjust RB and find the max and min values for I_E, I_B, V_E, and V_B as you did previously.

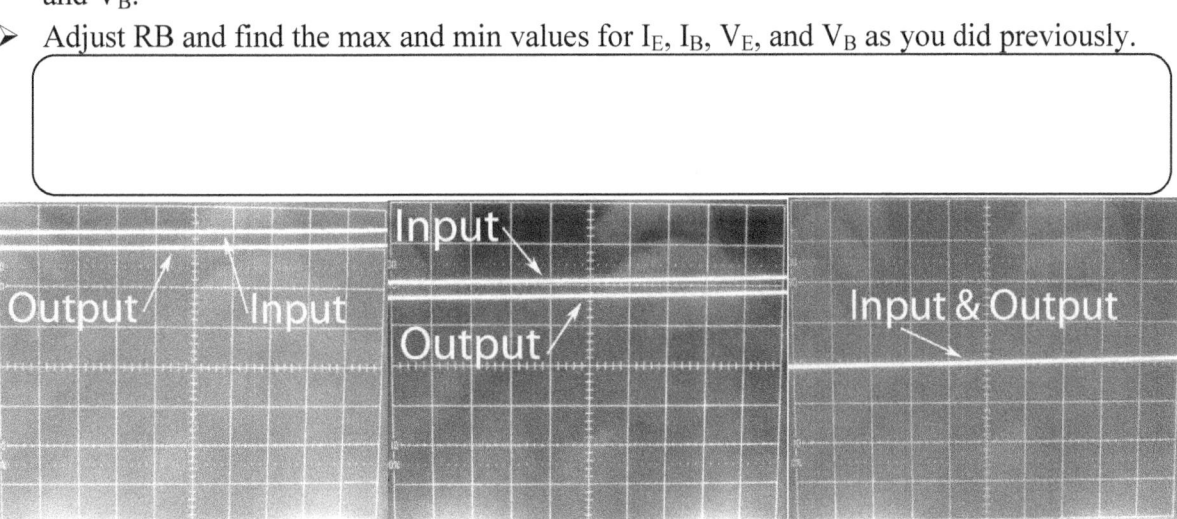

Figure 4. 9: The progression of the output and input voltages from the common collector as the potentiometer R_B is adjusted. Maximum (left), Intermediate (middle), Minimum (right). Note that the Input and Output signals are in 2Volt scale.

- As you adjust R_B, measure I_E, I_B, V_E, and V_B. Remember to keep your measurements in the ranges you just found.
- Make a graph of I_E vs. I_B to find β (Equation (4.1)). Based on your answer, is the common collector amplifier a good current amplifier?
- Make a graph of V_E to V_B. Based on your data, is the common collector amplifier a good voltage amplifier?

Part Ic: Common Emitter

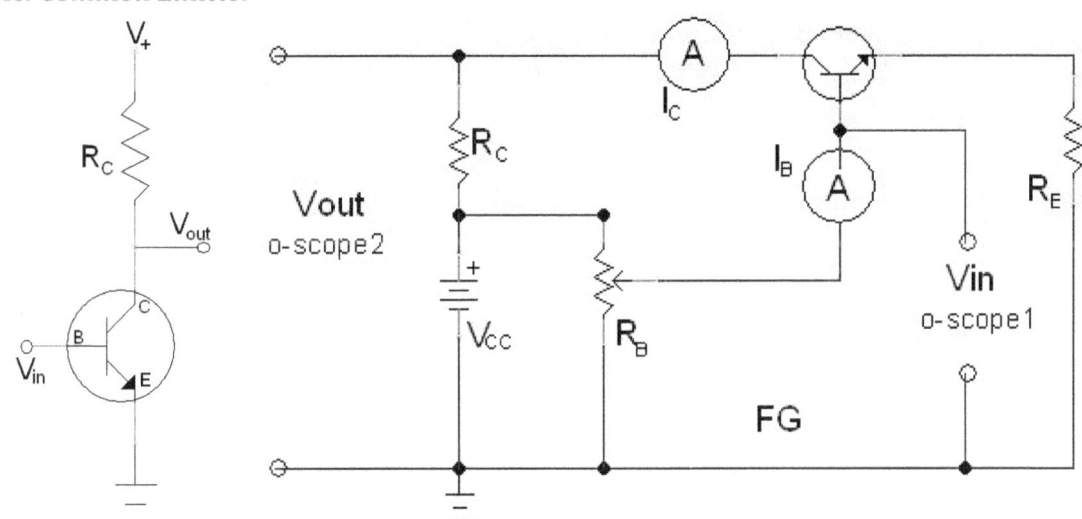

Figure 4. 10: The Common Emitter Amplifier. Top left) Simplest circuit design with no consideration of biasing. (right) A functional circuit diagram. Note that in an AC amplifier, the FG would be placed between the ammeter and ground.

The common emitter is so called because both the input signal and output signal share a common point at the emitter side. An input signal feeds a small current into the base of the transistor, which acts like a control valve for the current from V_{CC}. Input is measured from the base, while output is measured from the collector side of the transistor.

➢ Setup the circuit shown in Figure 4. 10, where R_E = 1kΩ, R_C = 5kΩ, and V_{CC} is +12V. R_B is a 500kΩ potentiometer used as a voltage divider. Refer to the Equipment section on how to use it if you are confused.
➢ Connect the DMMs to measure the currents I_E and I_B as well as the o-scope to measure V_C and V_B.
➢ Adjust R_B and find the max and min values for I_C, I_B, V_C, and V_B as you did previously.

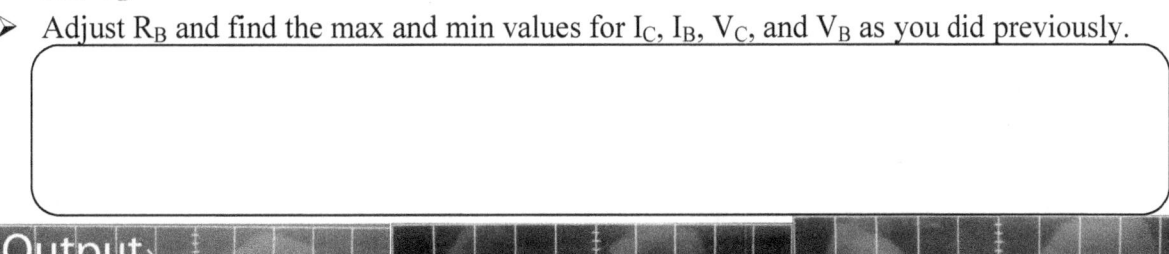

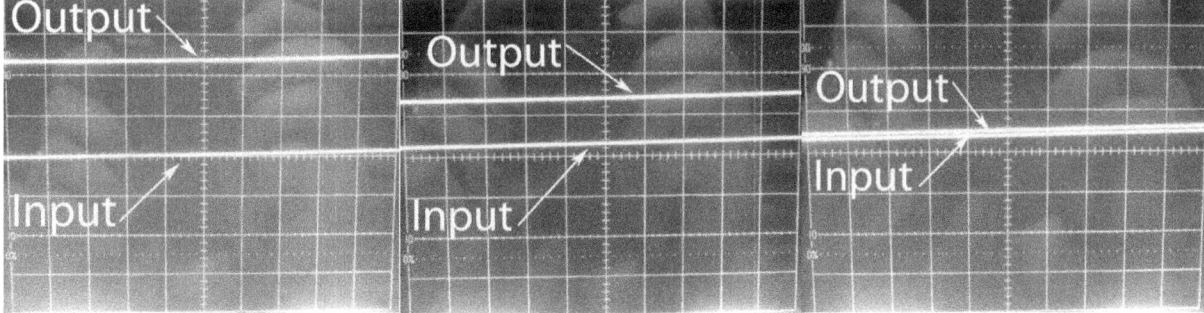

Figure 4. 11: The progression of the output and input voltages from the common emitter as the potentiometer R_B is adjusted. Maximum (left), Intermediate (middle), Minimum (right). Note that both the Input and Output is on a 5Volt scale.

- As you adjust R_B, measure I_C, I_B, V_C, and V_B. Remember to keep your current in the range of $1\mu A < I_B < 8\mu A$ and your voltage in a range of $0V < V_B < 1.5V$.
- Make a graph of I_C vs. I_B to find β (Equation (4. 1)). Based on your answer, is the common emitter amplifier a good current amplifier?
- Make a graph of V_C to V_B. Based on your data, is the common emitter amplifier a good voltage amplifier?

- Based on your answers, explain why the most common configuration (no pun intended) used in the real world is the common emitter.

Part Id: Measuring Transistors with the DMM

You will need to confirm the values of the gain you just calculated. Fortunately, your DMM has a function for measuring β. To do so, you will first need to remove the probes and plug in the rectangular shaped piece as shown in Figure 4. 12.

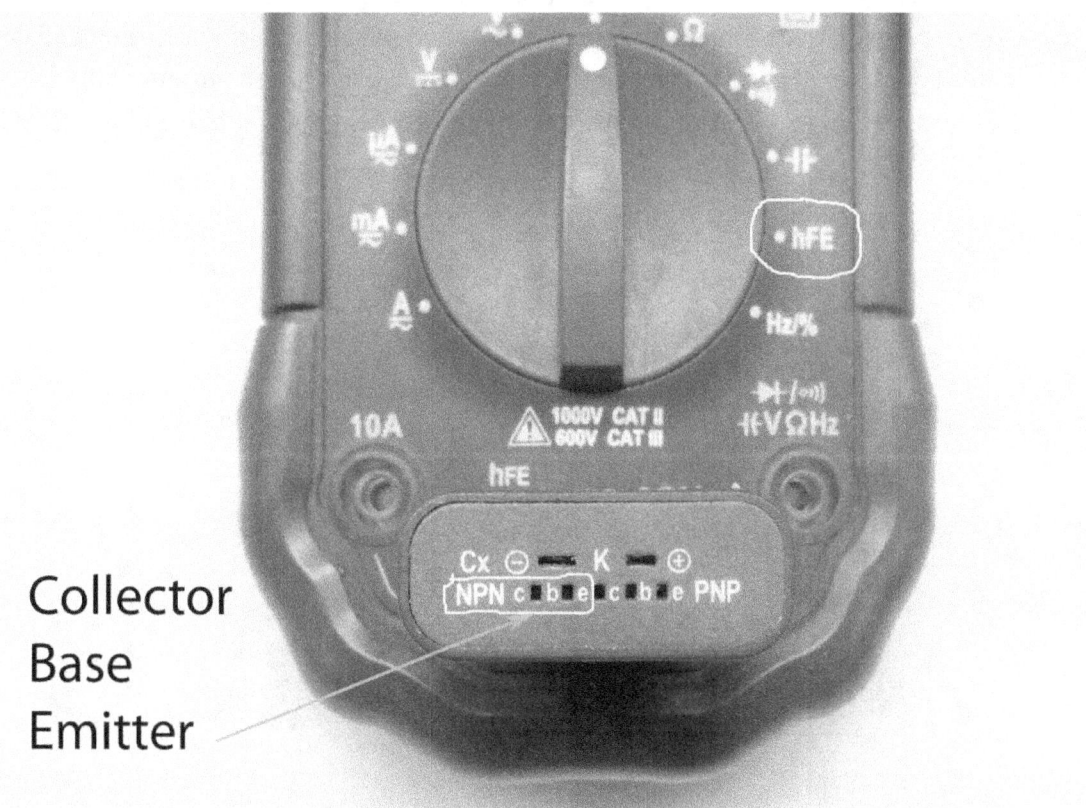

Figure 4. 12: The attachment for the DMM that measures the transistor's gain. Place the pins of the transistor in the holes for the collector, base, and emitter and move the knob to select hFE.

When the transistor has been properly placed, the DMM will give you the value of β. If no value is shown, or the value is single digits, either the transistor has been placed in the DMM incorrectly, or the transistor has been damaged. Use the value you get from the DMM as the "correct" answer and calculate a percent difference between this value and one you found in the Common Base.

Part II: AC Amplifier

Part IIa: Introduction

In the real world, AC signals are difficult to create with large currents while DC can have much higher currents much easier. As we have seen, a transistor can be used to amplify the current and is often used to amplify a weak AC signal. We found previously that the common emitter is the best setup to use for both current and voltage amplification, but we have yet to discuss how to properly bias the transistor to keep it firmly in the operational range of even while adding an AC signal.

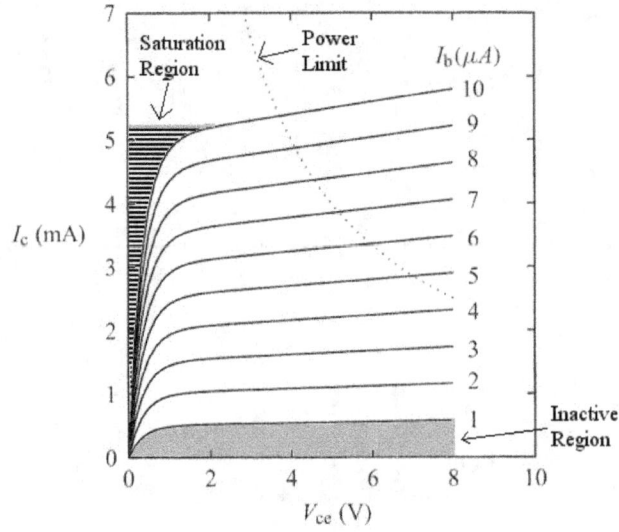

Figure 4. 13: The characteristic I-V curve of a small signal transistor.

In Figure 4. 13 we see the characteristic I-V curve of a typical small signal transistor, like the 2N2222A we are using. This extremely important, if confusing, graph is what holds the key to constructing the appropriate circuit for signal amplification. Failure to do so will result in the signal saturating, the transistor becoming inactive, or possibly even destroying the transistor itself.

We can see from the graph that the base current must be between $1 - 10\mu A$, while the voltage drop from the collector to the emitter should be between $1 - 8V$. This will yield a collector current of less than $10mA$. Should there be insufficient base voltage or base current, then the transistor will fail to activate (shaded region). If too much base current or too much collector current is supplied, then the transistor will saturate (lined region). Finally, if you provide a combination of too much voltage across the transistor and too much base or collector currents, then you have supplied too much power and the transistor will fail (dotted line).

Part IIb: Calculations for DC Amplification

Knowing all of this, we can start to design our AC signal amplifier.

The final circuit that you will build is in Figure 4. 20 and is a common emitter like the one you just built in the previous section. This design uses a voltage divider bias method that you have used in the previous setup. Not only is there no need for two DC power supplies, but it is the most widely used design and will likely be the kind you will use or see in the future.

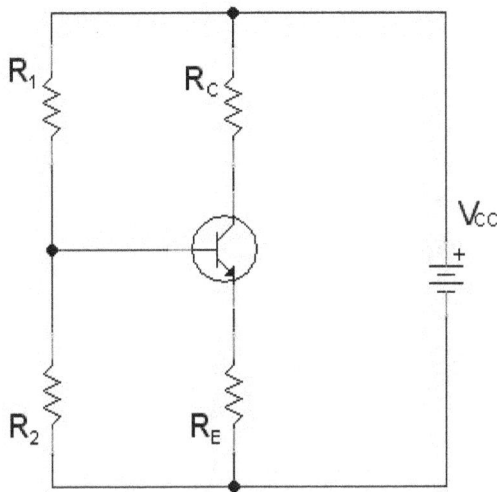

Figure 4. 14: The DC amplifier with a voltage divider base. Notice that this is a necessary first step toward building the AC amplifier.

We will build this circuit in stages to understand each part. Look at Figure 4. 14. This is functionally the same circuit as the common emitter you just built, but has been modified to have a voltage divider instead of the potentiometer you previously used. The values you will calculate in the following steps represent the estimates based on a generic 2N2222A transistor. Each transistor, however, will require slightly different values. Part IIc will allow you to fine tune your values.

We know the transistor needs at least 0.7 volts at the base to activate. To make the calculations easy, we can assume that Vcc = 12V, and the voltage divider will provide 3V of that voltage to the base. Using your equation for the voltage divider from previous labs, determine the relationship between R_1 and R_2 for $V_{out} = \frac{1}{4} V_{in}$.

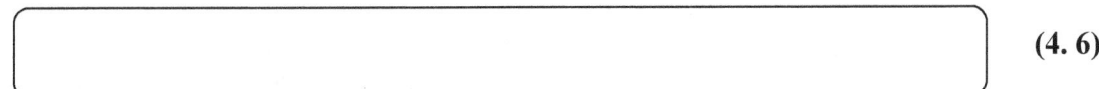 (4. 6)

From Figure 4. 13, we can tell that we want only a small current, less than $10\mu A$, running to the base to prevent saturation. Use Ohm's law to write an equation that relates the voltage through R_1 to the resistance of R_1 and a maximum current of $7\mu A$.

 (4. 7)

And then calculate the value of R_1.

Using that value and Equation (4. 6), calculate the value of R_2. Remember that the bias need only be more than 0.7V. If you cannot find the precise values you have calculated for your resistors, choose values that give an output voltage a little lower than 3.0V.

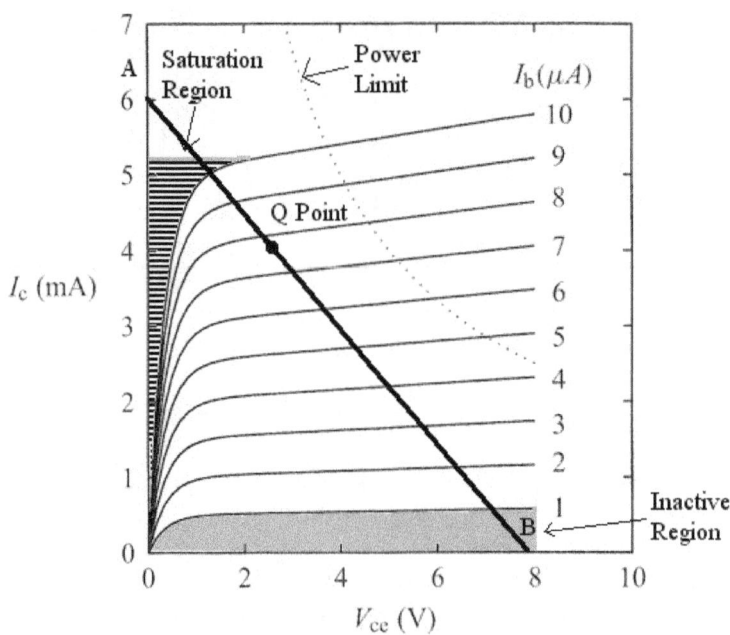

Figure 4. 15: The characteristic I-V curve for the 2N2222A transistor. The Q-point and the endpoints of the line one which the Q-point lies are indicated as A and B.

Referring to Figure 4. 15, we need to find the Q point (optimal operational point) for the transistor. We do this by finding the two points given by Equation (4. 8) (A & B) and the boundary conditions at those points.

$$I_C = \frac{V_{CC} - V_{CE}}{R_C} \qquad (4.8)$$

At the point A, you can see that $V_{CE} = 0$. At point B, you can see that $I_C = 0$. Write the equations for I_C at the points A and B below. These are possible conditions for saturation and cutoff respectively.

$$\qquad (4.9)$$

$$\qquad (4.10)$$

We can further see that the Q point would be at a value for V_{CE} halfway between V_{CC} and zero. At Q point:

$$V_{CE} = \frac{1}{2}V_{CC} \qquad (4.11)$$

If the potential difference across the transistor is given by Equation (4. 11), then the other half of the voltage must be across R_C. Look at Figure 4. 15, we can see that a safe value for the collector current is ~4mA. Knowing that the current through R_C must be the same as the collector current, use Ohm's law to calculate the value of R_C.

$$\boxed{} \qquad (4.12)$$

We know for transistors that

$$I_E = I_C + I_B \qquad (4.13)$$

But, $I_B \ll I_C$, so we can say $I_E \approx I_C$. We can now apply Ohm's law again to find R_E, where the voltage across R_E is V_{CE}.

$$\boxed{}$$

A good check on your values is to use Equation (4. 14) with the gain approximated to 200.

$$R1 || R2 < \beta Re \qquad (4.14)$$

- Build the circuit with the values you just calculated. If you find that the voltage to the base is less than the activation of the transistor (~0.7V), then choose a lower value for R1 and try again. Shoot for a voltage to the base of about 2 Volts.
- Calculate the value of β using the same technique as you used in Part I for the Common Emitter (just one data point will do). Check that the value of β is the same as you found before.

Part IIc: Simple AC Amplification

Despite all that work, we have only built another DC amplifier like the one you built in the last section and the values of all the components are only approximately correct. We now need to add the AC signal to our setup and find the exact values that place the circuit in the optimal operational position on the I-V graph. Examine Figure 4. 16; the potential AC signal has been overlaid on the graph at the Q point. We can easily see that if the amplitude is too high or the Q-point is misplaced, the circuit may transition into either the saturation, inactive, or power limit regions.

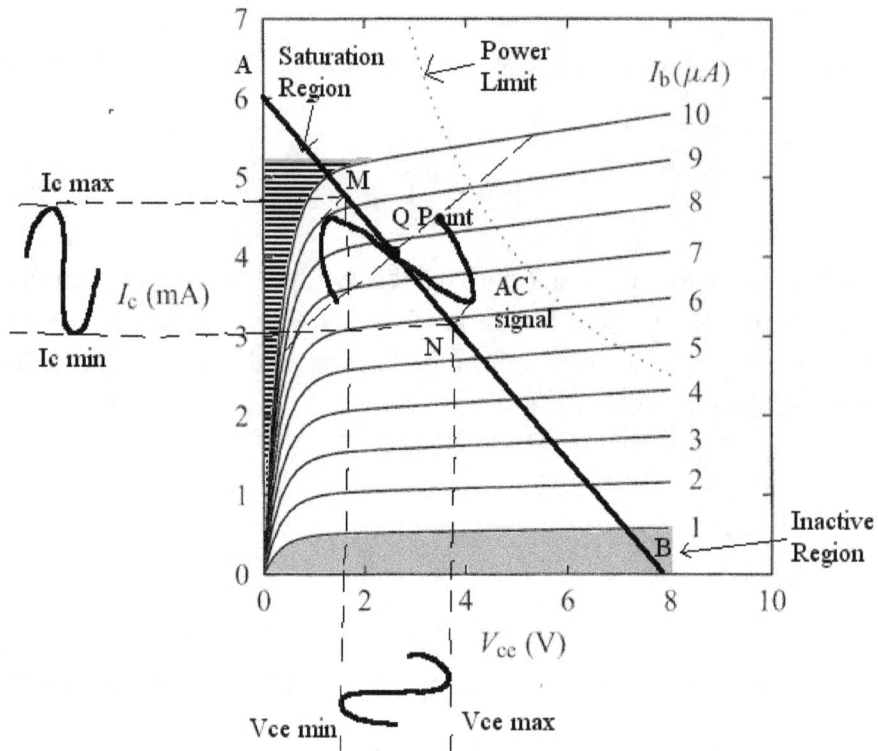

Figure 4. 16: The I-V characteristic curve with AC signal overlaid. Notice how, if expanded, the AC signal could run into the saturation region or the power limit.

➤ Set the FG to a V_{rms} of $2.0V$ as measured by your DMM. From Figure 4. 17, we can see that the FG will feed directly into the base of the transistor. We further know that the base current must stay below $10\mu A$. Use Ohm's law and calculate the value of R_L necessary to bring the current from the FG down to 5 μA.

While we have done everything we can to peg this circuit to a good Q-point, every transistor is different. We must treat the values you have calculated as the starting points for a well turned circuit. Examine Figure 4. 17.

➤ Replace R_1 and R_2 voltage divider with a 500kΩ potentiometer used as a voltage divider.
➤ Replace R_C with a 10kΩ potentiometer used as a variable resistor.
➤ Replace R_L with a 500kΩ potentiometer used as a variable resistor.
➤ Hook up the FG where shown in Figure 4. 17 and set it to $1000Hz$ and keep the amplitude very low. Examine the output and input signals on the o-scope set to DC.

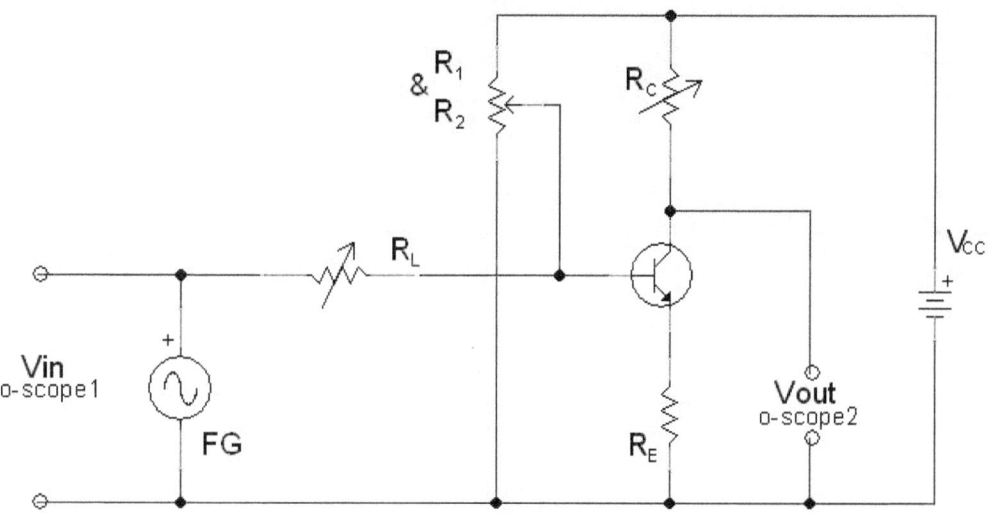

Figure 4. 17: The AC amplifier. This is just the DC amplifier with an AC signal fed into the voltage divider and to the base of the transistor.

- Adjust the three pots and the amplitude of the FG until the waveform of the output is a clean sine wave with no clipping (Figure 4. 18). Take care that you do not turn the pots to zero resistance as you may destroy the transistor!
- Get the largest amplitude increase possible with no distortion and take a picture. If there is distortion, you have most likely not set the Q-point in the center of the operational range!
- Measure the resistance of each potentiometer (R_1, R_2, R_C, and R_L) so you know in the future what values are needed for this transistor.

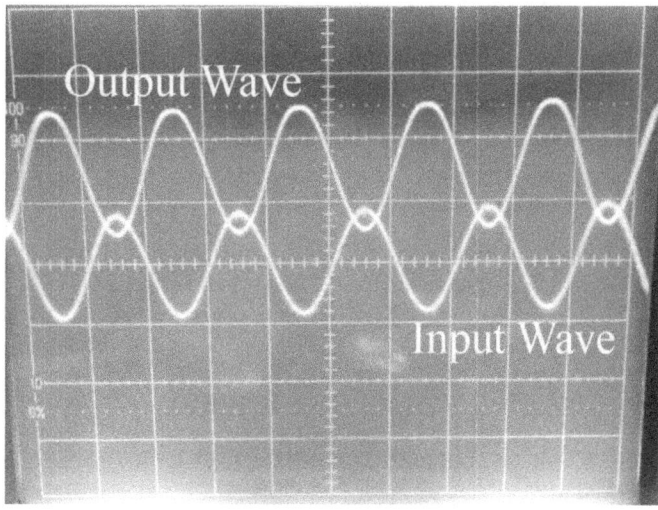

Figure 4. 18: The Common Emitter input and output waves with the Q-point set to the middle of the operational region on the I-V curve.

Part IId: High Pass Filter

Often, the common emitter amplifier incorporates a filter or two to limit the frequencies over which the AC signal will be amplified. We will now add a high pass filter to our amplifier.

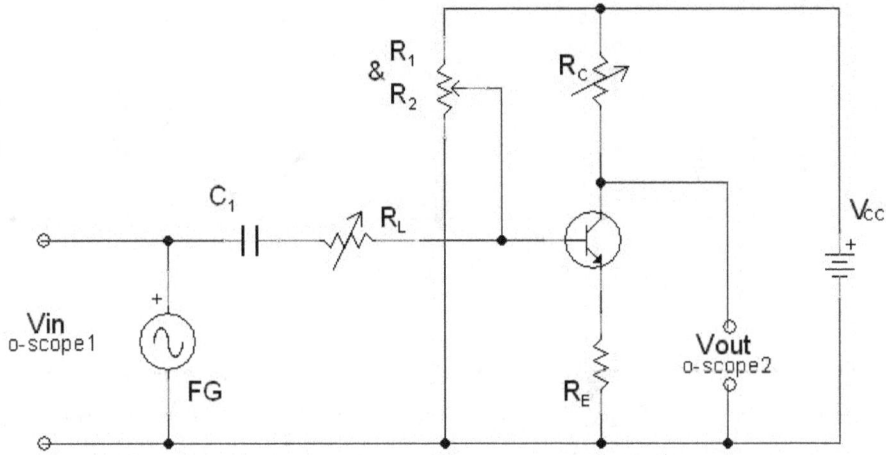

Figure 4. 19: The common emitter circuit with a high pass filter.

➢ It is difficult to see which resistor is combined with C_1 to make the high pass filter. If you use your imagination, you can see that R_2 is the resistor that forms the high pass filter. This will be one side of the potentiometer.

➢ A capacitor will be added as shown in Figure 4. 19. Calculate the value of capacitor needed, in combination with R_2, to filter out all frequencies less than $400 Hz$.

➢ Place the capacitor at the position of C_1 and observe the results.

Part IIe: Removal of DC Bias

A second capacitor can be used to eliminate the DC bias from the output. At higher frequencies, C_2 will act as a short to ground, allowing the AC signal to pass through. At low frequency (the DC current), the capacitor will act as an open circuit, preventing the DC bias from passing to ground. Since the low frequencies where filtered out in the previous step, we don't have to worry this capacitor affecting the FG input signal.

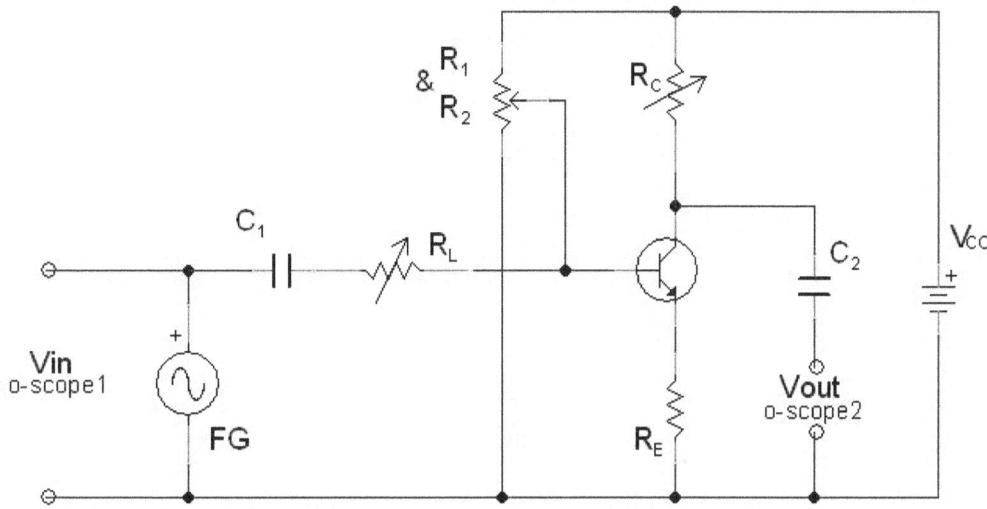

Figure 4. 20: The final version of the AC amplifier, complete with a second capacitor to filter out the DC bias signal.

- Choose ceramic capacitor for C_2 with a value greater than $0.1\mu F$ (Figure 4. 20), the bigger the better.

- The capacitor will charge from the DC signal, but, because of the large capacitance, allows high frequency (AC) to pass through. If you turn on the power first, and then connect the o-scope probe (set to DC), you can see the signal drop down to zero on the screen.
- If you miss the movement, remove the capacitor from the circuit, discharge it on some piece of metal and then re-insert it into the circuit to try again.
- Record the output.

Part III: Transistors as a switch

The final stage for the Common Emitter, Voltage Divider Base, AC amplifier you've built, is to put it to work as a switch. We are going to control a LED with this circuit by changing the various potentiometers until the transistor slides into the saturated region of its I-V curve.

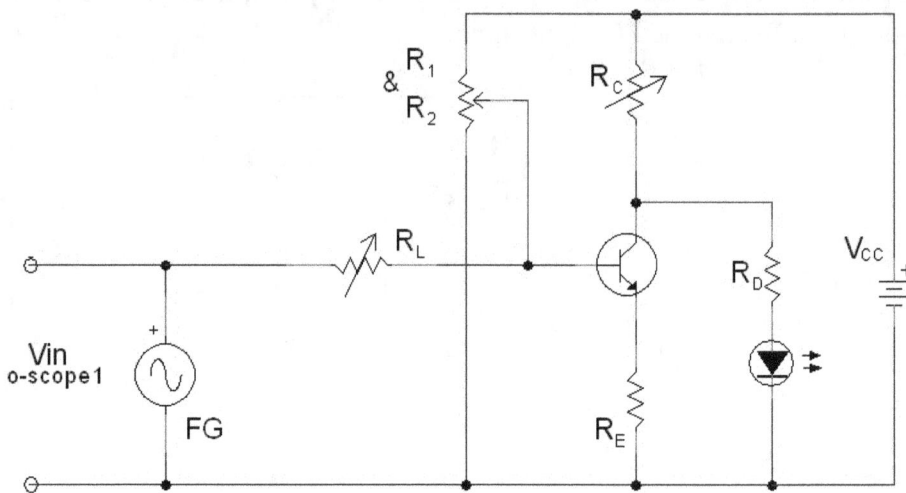

Figure 4. 21: Transistor used for switching on the LED.

We can now make the transistor do some switching.

- Remove both capacitors from the circuit. We are going to be using very low frequency to see a blinking LED, filtering out low frequencies would be counterproductive.
- Adjust the potentiometers until the top of the output wave (and only the top) is clipped (Figure 4. 22). Take a picture. Measure and record the resistance of the potentiometer you changed.

- Explain in your lab, using the Q-point, why it is clipped off and how changing the potentiometer you changed achieved the outcome.
- Adjust the potentiometers until you clip the bottom of the output waveform (and only the bottom). Make sure that the clipped portion is below 1V. If the clipped portion will not go below 1V, add a 5kΩ resistor in series with R_C. Take a picture.
- Explain in your lab report using the Q-point figure, why this occurs, what you changed to make it occur, and how the changes achieved the outcome
- Remove the o-scope probe, and replace it with an LED connected to ground and a load resistor in series, $R_D = 20Ω$ ½ Watt. Lower the frequency to less than 30 Hz, and observe the LED. You should see the LED blinking. Try adjusting R_C while the LED is blinking. How does the blinking change? Explain why the blinking changes the way it does.

Note that the LED could be replaced with a motor, and you could use this technique to control the motor speed.

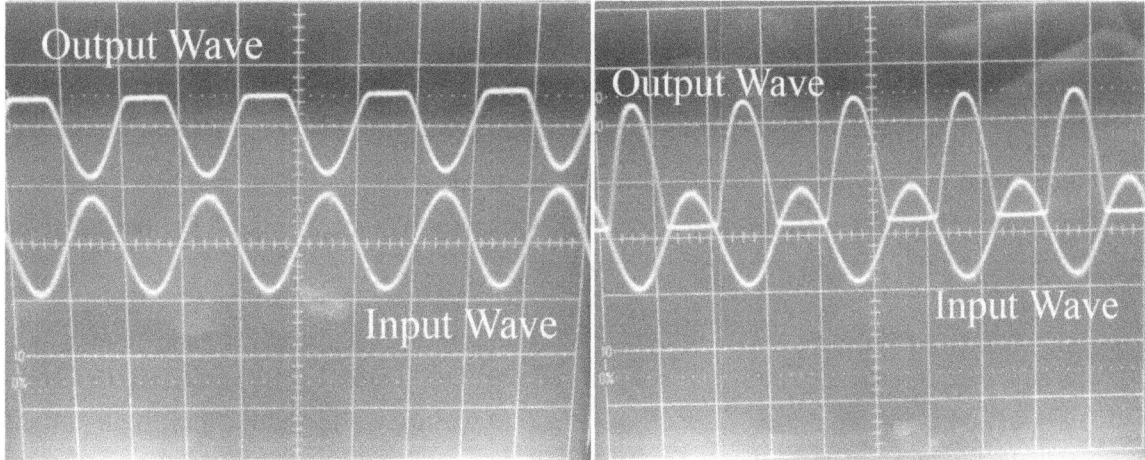

Figure 4. 22: The output waveform when the Q-point overlaps with (left) inactive region or (right) saturation region.

Summary

In this lab, you have examined some of the properties of transistors. You have also investigated some of the uses of transistors in simple circuits. In particular, you have used transistors to make a simple emitter follower (near unity voltage but with current gain) and an inverting amplifier (voltage gain and inverting).

Part IV: Further Applications

i. The Darlington Pair: The overall current gain is given by the gain of the first transistor multiplied by the gain of the second transistor as the current gains of the two transistors multiply. If two identical bipolar transistors are used to make a single Darlington device then the overall current gain will be given as β^2.

This pair does not need to be built each time; it comes pre-packaged in what looks like a large single transistor. The two most common of these is the TIP120 and TIP102. These two handle 60W and 120W respectively making them ideal for controlling high amperage devices like motors and solenoids.

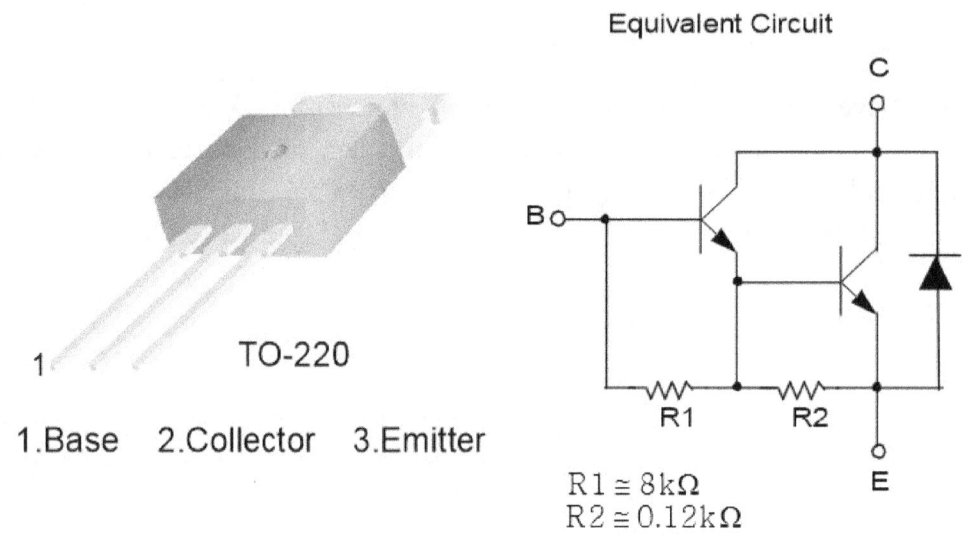

Figure 4. 23: The Darlington Pair of transistors. This is often packaged in one piece, appearing to be just one transistor as shown here as the TIP120. Notice the diode across the collector to emitter junction; it prevents voltage spikes going back through your circuit and damaging your microprocessor.

Part V: Prelab Questions

(1) In Part I, you will be building the Common Base, Collector, and Emitter configurations. For each configuration, find current gain and voltage gain for each configuration. Write both the equation, and their approximate value (<1, ~1, >1). During lab, you will use this table to make sure your circuits are working properly.

(2) In Part II (a-c), there is a mostly complete set of instructions on how to calculate the appropriate values for the components for an AC amplifier. Work through the instructions and fill in the empty boxes BEFORE you come to lab. Failure to find the values asked for will dramatically slow down your group.

Lab 5 - Introduction to Operational Amplifiers

Goals: To study common uses for the Op Amp including:

1. Comparator
2. Unity Gain Buffer
3. Non-Inverting Amplifier
4. Inverting Amplifier
5. Integrator
6. Differentiator

List of Equipment and Parts

1. Oscilloscope
2. Function Generator
3. DC Power Supply
4. 2 Digital Multimeters
5. Solderless Breadboard
6. 1 Capacitor (0.01µf)
7. 12 Resistors (100Ω, 400Ω, two 1kΩ, 4.3kΩ, 5kΩ, two 10kΩ, 16kΩ, 20kΩ, 850kΩ, 1MΩ)
8. 1 Op Amp (Quad 741)
9. Jumper Wires
10. 3 Banana wires
11. Alligator clips

Introduction and Background

Negative feedback sounds bad, but in control systems, it gives stability. Consider a simple control system, the thermostat in a room. When the temperature goes down, the thermostat sends a negative feedback signal to the furnace, which makes the temperature go back up. If the feedback were positive, an increase in temperature would lead to a further increase in temperature until something gave out (e.g. the boiler burst) or until the system hit a limit (the temperature would level off when the house dissipated as much heat as was being put into it). You may remember that the voltage regulation by the ML317 chip in your low voltage power supply results from negative feedback.

GOLDEN RULES of OpAmps:

I. **Op amp inputs draw no current** (this is a consequence of the amp's high input impedance; in the ideal case, we assume this impedance is infinite).

II. V_{out} **will try to adjust so that** $V- = V+$. (this is a consequence of the amp's high gain; in the ideal case, we assume this gain is infinite. In reality, the two inputs are not quite driven to equal voltages; instead, $V^- - V^+$ is small, typically microvolts.

Note: the names and labels for the two input terminals, $V-$ and $V+$ sometimes cause confusion. Please notice that these terminals are *not(!)* power supply terminals. Power supply connections usually are *not shown* on diagrams of operational-amplifier circuits; but the op amp's power supplies must always be driven. This connection is omitted from the drawings because it is thought to go without saying. They are not indications of a secret battery or battery-equivalent that lies within the integrated circuit!

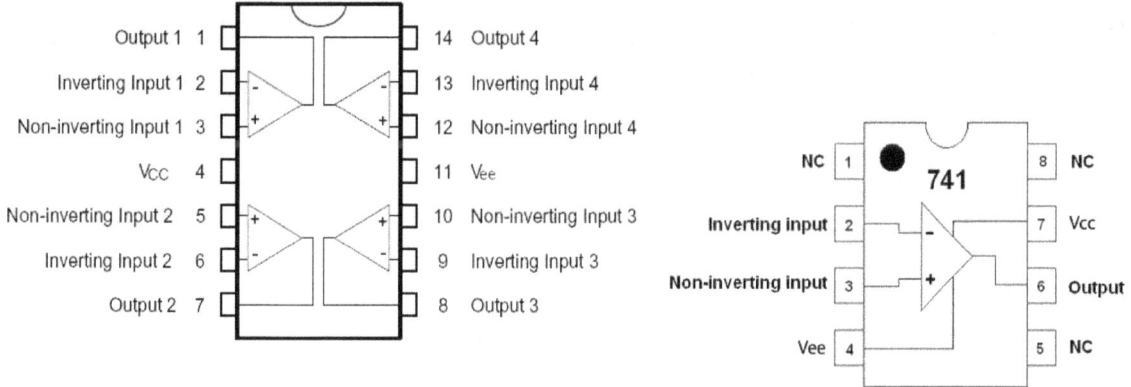

Figure 5. 1: Diagrams of the two most used 741 configurations; Quad (left), Single (right).

Procedure: Basic Op-Amp Characteristics

Figure 5. 2: Placement of the Op Amp on the breadboard

We will be using the Quad 741 op-amp for this lab (Figure 5. 1), which is really four op-amps in one. The op amp must be placed straddling the "seam" of the breadboard so that none of the pins are directly connected to another (Figure 5. 2). Note that the V_{CC} and V_{EE} will be the +12 V and -12 V from the power supply.

Part I: Comparator

When Op amps amplify a signal, the output signal must swing quickly for a small change in the input signal. The speed of this response is called the slew rate and is measured in Volts/microsecond.

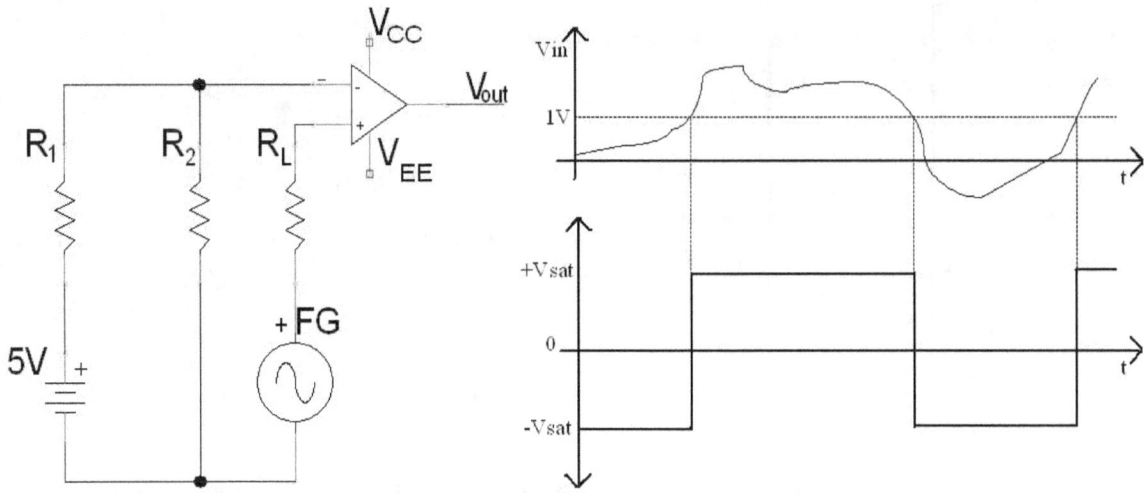

Figure 5. 3: The Comparator circuit and its response

To measure the slew rate, we will be building the comparator circuit shown in Figure 5. 3. The comparator compares an input signal to a reference signal. If the input signal is above the reference, the Op amp amplifies the signal with an infinite gain. The output signal is amplified until it reaches the saturation level of V_{CC}. If the input signal is below the reference signal, then the output signal swings to V_{EE}. Because the amplification is limited in speed by only the Op amp and no external factors, this circuit allows us to easily measure the slew rate.

➢ Build this circuit and use the FG to supply a sine wave to V_{in}. Use the +12V and -12V to supply voltage to V_{CC} and V_{EE} respectively. It is important that you do not switch polarity of V_{CC} and V_{EE} as it can destroy the op-amp. The output from a voltage divider, fed by a 5V input, will supply the reference voltage of ~1V. Finally, be CERTAIN to use resistor values for R_L and R_1 of sufficiently high value to lower the input current to between 1 and 10 mA.

➢ Use the o-scope to chop between both the input signal and the output signal as you vary the amplitude of the FG sine wave. Adjust the FG until the amplitude of the input wave is less than the reference voltage. Take a picture or record what you see on the o-scope and explain why the outcome is what you see.

➢ Adjust the input wave until the amplitude is greater than the reference voltage and record what you see on the screen. Again, explain in your lab report why the output is this form.

- Change V_{CC} from +12V to +5V and apply the same amplitude sine wave you have before. Record the o-scope and explain why it looks the way it does.

- Change V_{CC} back to +12V and switch the FG to a square wave. Make a prediction about how the output should look and then observe the outcome. Is there a difference between your prediction and observation?

- Notice that the comparator circuit can use any voltage as a reference signal, even zero volts. Remove the 5V input from the voltage divider and connect the inverting pin to ground (reference voltage) and observe the affect it has on your outcome

- Reconnect the voltage divider to the inverting pin so that the reference voltage is once again positive. Importantly, we can use the comparator to determine the slew rate, a measure of how fast (V/sec) the 741 Op-amp can respond. Leave the FG on square wave and increase the frequency until the vertical section of the square wave is noticeably no longer vertical. Measure this slope and record it as your slew rate (V/μs). You will use the slew rate to answer questions later.

- Finally, if you keep increasing the frequency above where you measured the slew rate, you will start to see a triangle or an inverse triangle wave. Explain in your lab report using the slew rate why the output wave looks this way.

Part II: The Unity Gain Buffer

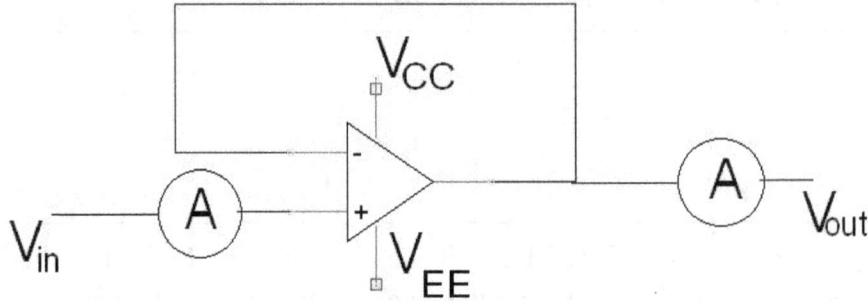

Figure 5. 4: Unity Gain Buffer

This circuit is often used when a voltage source with a high internal impedance is used, and you want to draw more current than the source can deliver. The solution is to use this circuit to make a copy of the original voltage, V_{in}, and that copy, V_{out}, appears at the output of an op-amp. The V_{out} of the unity gain buffer can often deliver more current without lowering the output voltage because the internal resistance of the op-amp is lower than the internal resistance of the original source.

➢ Now we will build the Unity Gain Buffer in Figure 5. 4, where V_{in} is a sine wave input from the FG and V_{CC} and V_{EE} are the same as in the previous section. Previously, we added a resistor in series with the FG to keep the current low, but the resistors in the ammeters used to measure the input and output currents will keep the current low enough.

➢ Measure the V_{in} and V_{out} to confirm that there is no gain.

➢ Also, set your ammeter to measure AC current and <u>quickly</u> measure I_{in} vs. I_{out}. Does the current gain change with amplitude?

Part III: The Non-inverting Amplifier

The Non-Inverting Amplifier shown in Figure 5.5 is simply the unity gain buffer with two added resistors. These resistors allow you to adjust the voltage gain of from the V_{in} to the V_{out}. We can determine the equation for this gain by noting that the current rule will force the current to the inverting terminal (-) to be zero. Also, according to the second golden rule, the voltage at the inverting terminal needs to match the voltage at the non-inverting terminal (+). These two conditions give a node equation of

$$\frac{V_{in}}{R_1} = \frac{V_{out} - V_{in}}{R_f} \qquad (5.1)$$

Solving, the transfer function can be found to be

$$A = \frac{V_{out}}{V_{in}} = 1 + \frac{R_f}{R_{in}} \qquad (5.2)$$

Where A is the theoretical gain.

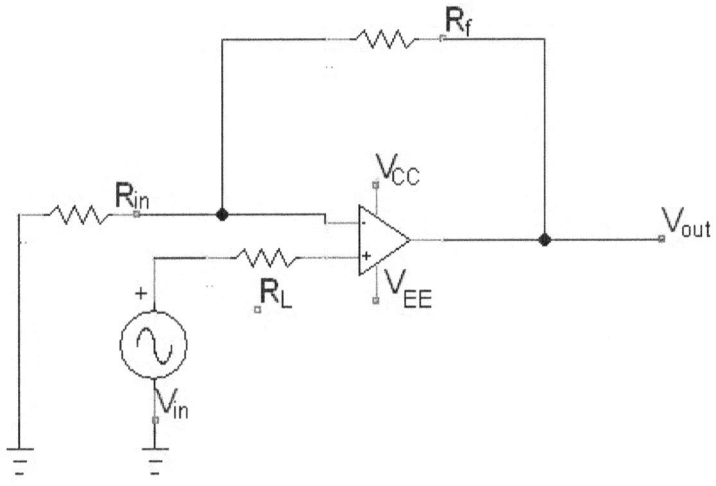

Figure 5. 5: Non-Inverting Voltage Amplifier

- Choose the necessary resistor values to yield a theoretical gain of 2.

- Build the non-Inverting Amplifier in Figure 5. 5. As with the unity gain buffer, the FG will be the V_{in}. Be sure to add a current limiting resistor in series with the FG to keep the current low, just as you have done previously.
- Use the o-scope to examine both the V_{in} and V_{out} to measure the gain. Does the gain match your theoretical gain? If not, what assumption did we make about the Op-amp that could change the gain?

- Increase the FG amplitude until the output wave deforms. Explain in your lab report what causes the wave to deform, and three quantities you could alter to keep the output wave from deforming.

- Increase and decrease the frequency while recording the gain at the full range of the FG. (No need to take many data points until the gain starts to change). Make a plot of log gain (dB) vs log frequency. Does the measured gain remain constant at all frequencies?

- Explain in your lab report using the slew rate why and at what frequency the wave deforms.

Part IV: The Inverting Amplifier

The Inverting Amplifier is so called because the output is 180 degrees out of phase of the input. The input terminals need to have zero difference between them, so there has to be zero volts at the inverting terminal (-) due to the fact that the non-inverting terminal (+) is grounded. This leads to the node equation

$$\frac{V_{in}}{R_{in}} = \frac{-V_{out}}{R_f} \qquad (5.3)$$

Solving, the transfer function gives the amplitude gain,

$$A = \frac{V_{out}}{V_{in}} = \frac{-R_f}{R_{in}} \qquad (5.4)$$

Where A is the theoretical gain. Notice that if $R_{in} = R_f$, this will be a unity gain inverter.

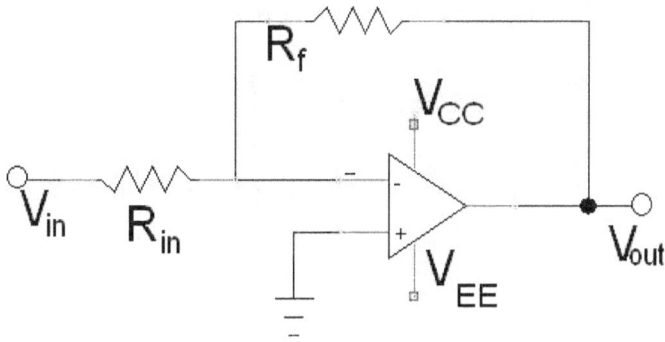

Figure 5. 6: Inverting Voltage Amplifier diagram

- Choose two 1% quality resistors of approximately 10K ohm for R_f and R_{in}, and build the circuit in Figure 5. 6. Use the FG for V_{in} and set the o-scope to view both V_{in} and V_{out}, and record the output. Does the outcome match the predictions from theory in Equation (5. 4)?

- Switch out one of the resistors to obtain a gain of 2, and repeat the same steps from the non-inverting amplifier above to measure the frequency response of the circuit.

- Finally, set the gain to 100, the amplitude on the FG to 1 V and observe the output when the op amp saturates.

Part V: Integrating Circuits

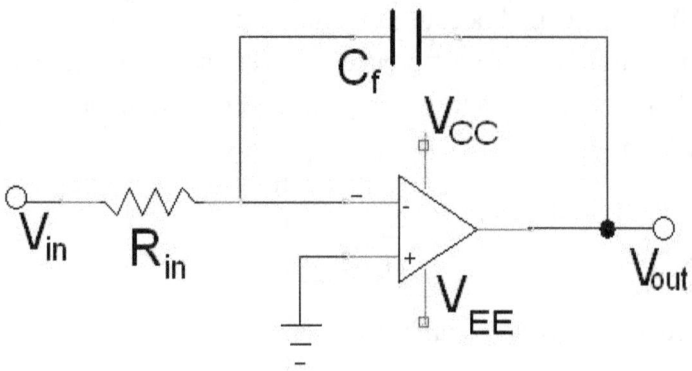

Figure 5. 7: The Ideal Integrator Circuit

If we take the inverting amplifier and replace the R_f with a capacitor C_f, the current through the output will now flow into C_f which will begin to build up a charge. As the charge increases, the voltage across C_f will then increase at a rate proportional to the current.

$$I_{out} = \frac{dQ}{dt} = -C_f \frac{dV_{out}}{dt} \tag{5.5}$$

From the Op Amp rules, the current through R_{in} is equal to the current through C_f, which when combined with Equation (5. 5) we get,

$$I_{in} = \frac{V_{in}}{R_{in}} = -C_f \frac{dV_{out}}{dt} \tag{5.6}$$

If we rearrange the above equation and solve for V_{out}, we get the equation for the Ideal Integrator.

$$V_{out} = -\frac{1}{R_{in}C_f} \int V_{in} dt \tag{5.7}$$

We can see from this equation the output voltage is an integral of the input voltage but with a weighting equal to $-\tau^{-1}$. This weight will affect the slope of the integrated output signal which will, in turn, affect the amplitude of the output wave. Also, the negative will invert the waveform, which should be no surprise as this is just a modified inverting amplifier.

Notice that the *dt* in the integral will change with frequency, altering the DC gain. This should also come as no surprise, since we have just built a low pass filter to the inverting amplifier. The magnitude of the DC gain is the same as for an inverting amplifier (Equation (5. 4)), but with the reactance (χ) of the capacitor instead of R_f.

$$|A| = \frac{V_O}{V_{in}} = \frac{-X_c}{R_{in}} = \frac{\left(\frac{-1}{2\pi f C_f}\right)}{R_{in}} = \frac{-1}{2\pi f R_{in} C_f} \qquad (5.8)$$

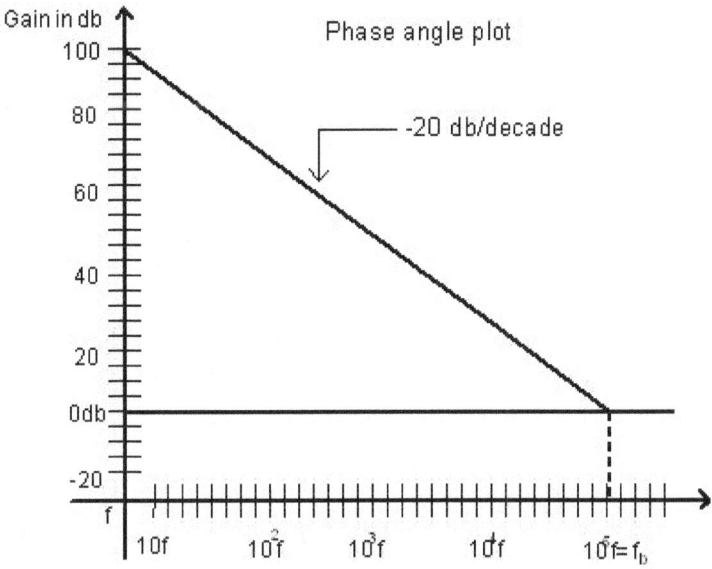

Figure 5. 8: Ideal Integrator Frequency Response. Notice how, as the frequency drops, the gain becomes very large.

The ideal integrator in Figure 5. 7 is, however, not useful for practical applications. As can be seen in Equation (5. 8) and Figure 5. 8, in the absence of input voltage or for $f \cong 0$ (D.C.), both the AC and DC gains is extremely high. Thus, the output of an ideal integrator circuit is likely offset towards the positive or negative saturation levels. More importantly, the offset voltage and bias current will nearly always distort the output waveform beyond any use.

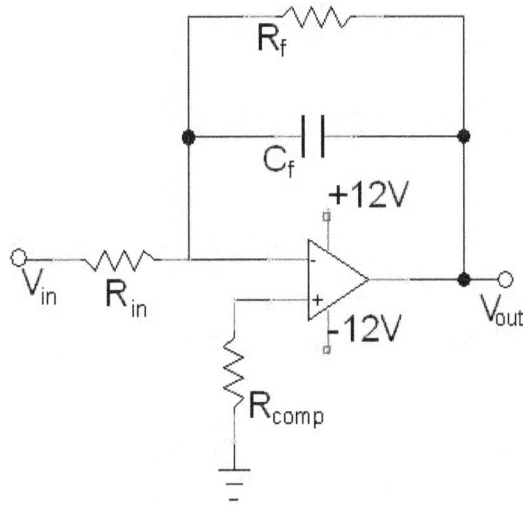

Figure 5. 9: The Practical Integrator. Notice the resistor in parallel with the capacitor.

We will have to modify the ideal integrator to make it work in a lab as shown in Figure 5. 9. The output voltage of the practical integrator is (after applying KCL)

$$V_O(s) = -\frac{1}{sR_{in}C_f + \frac{R_{in}}{R_f}} V_{in}(s) \qquad (5.9)$$

If R_f is very large then R_{in}/R_f become negligibly small, and the output voltage is the same as for the ideal integrator.

$$V_O(t) = -\frac{1}{R_{in}C_f} \int V_{in}(t) dt \qquad \text{where } \frac{1}{s} = \int dt \qquad (5.10)$$

If we reexamine the DC gain from Equation **(5. 8)** in the context of the new circuit, we can find the magnitude of the D.C. gain using the total impedance of R_f and C_f:

$$|A| = \frac{V_O}{V_{in}} = \frac{-Z}{R_{in}} = -\frac{(R_f \| X_c)}{R_{in}} = \frac{1}{R_{in}\sqrt{\left(\frac{1}{R_f}\right)^2 + (2\pi f C_f)^2}} \qquad (5.11)$$

Under DC conditions, the frequency in the above equation is zero, and the DC gain reduces to just the gain of an inverting amplifier (Equation **(5. 4)**). This is important, because, whereas the ideal integrator would have a varying DC gain, the practical ideal integrator can have a constant DC gain.

The practical integrator will not work at all frequencies, and care must be taken when designing the circuit so that integration occurs at the frequencies desired. The bounds for proper integration are determined by (at lower frequencies) where the output wave begins to deform, and (at higher frequencies) where the output wave's amplitude is no longer greater than the input amplitude.

The upper frequency limit, f_b, is determined by the low pass filter composed of R_{in} and C_f.

$$f_b = \frac{1}{2\pi R_{in} C_f} \qquad (5.12)$$

While integration still occurs above this frequency, other restrictions covered further on will keep the input signal at low amplitude, so an AC gain less than unity would lower that amplitude even further.

The lower frequency bound is typically defined as the break frequency, f_a. This frequency is the result of the deformation of the wave due to the RC time constant from R_f and C_f from the feedback voltage from V_{out} to the inverting input of the op amp.

$$f_a = \frac{1}{2\pi R_f C_f} \qquad (5.13)$$

While this frequency seems to be the accepted standard, significant wave deformation can be seen well above f_a. For precise results, this lab will use the lower limit of f_c.

$$f_c = \frac{1}{R_f C_f} \qquad (5.14)$$

This frequency is the point at which the RC circuit can only charge to ½ the time constant, keeping the output wave from being distorted.

Before designing the integrator, the above equations yield two principles that must be kept in mind:

1. A low DC gain is preferable for high amplitude inputs to prevent saturation. High DC gains will results in saturation of the output signal.
2. A high DC gain is preferable for large integration frequency ranges. Low DC gains will greatly limit the range over which the integration maintains its waveform.

These two principles mean that the integrator must be built with a specific range of frequencies in mind, or for a specific range of input voltages. Because amplitude of a wave can be modified with an amplifier without much difficulty, we will be building out integrator for a range of frequencies.

➢ We are now going to design a practical integrator circuit to integrate a square wave with a range of frequencies from $f_c = 10\ kHz$ to $f_b = 20\ kHz$. First, we want to protect the op-amp by limiting the current running into the input to a range of 1 to 10 mA. Use Ohm's law and your input voltage (~1V) to calculate an appropriate value for R_{in}.

➢ Use Equation (5. 12) and the value of R_{in} you just calculated to find the value of C_f.

➢ Now use the value of C_f and Equation **(5. 14)** to find the value of R_f.

➢ Finally, R_{comp} is equal to $R_{in} \| R_f$.

➢ Build the practical integrator using the values you just calculated. The range over which the circuit integrates properly is between f_c and f_b. Predict what wave form will be output when you integrate a square wave. Input a square wave of 1V and observe the output.

- Place the o-scope on AC for both input and output. Measure the peak voltage of the input wave and, using Equation **(5. 10)**, calculate the expected value of the slope of the output wave. Compare this value to the slope of the output wave.

- Switch the o-scope to display the output wave in DC. The waveform will be displaced from the zero line by some amount. Measure how large this displacement is from zero to the average line of the wave. Compare this value to the DC gain from the inverting amplifier, Equation **(5. 4)**.

- Increase the amplitude of the input wave until the output waveform begins to show saturation from the op amp. Explain in your lab report, using the DC gain, input amplitude, and output amplitude why saturation occurs at that amplitude. Also, explain why saturation is frequency dependent.

- You will need to make a graph for your report of $20 \log \left(\frac{V_{out}}{V_{in}}\right)$ as a function of frequency. Move the frequency from 100Hz to f_b, recording the ratio of the V_{out}/V_{in} along the way. To speed up the data collection, note that f_c to f_b is linear and 100Hz to f_a is linear in the log graph.

- Include on your graph what the shape of the output waveform is in each region and why it looks as it does.

- Note the noise level in the input signal from the FG and compare to the noise level after integration. Use your graph of amplitude vs. frequency to determine why high frequency noise is filtered out when integrating.

- Just as in the transistor AC amplifier you built, the DC gain needs to be removed so that the output signal is pure AC. Do this by adding a ~1μf capacitor to the circuit as you did in the transistor lab at the output (although do not connect the capacitor to ground). Check that the DC component of the signal has been removed without altering the output signal before moving on.

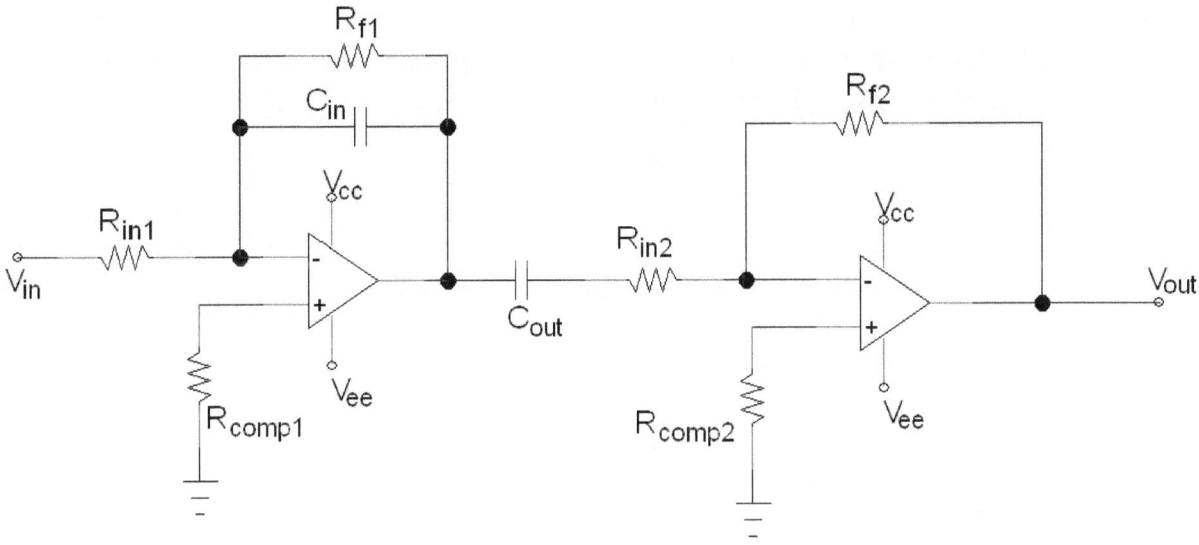

Figure 5.10: The practical integrator feeding an inverting amplifier. Because both invert the signal, the final output is non-inverted.

- Next, get a second 741 op amp and build the inverting amplifier with a gain of ~10. Feed the output wave from the integrator into an inverting amplifier. Examine the input and output to the amplifier and confirm that the gain is 10 and that the signal is no longer inverted.
- Note that the choice at the beginning to build the integrator to a specific range of frequencies has now paid off. Any loss in amplitude can always be regained through the addition of another amplifier.

Part VI Summary

This lab has explored several basic functions that can be performed with a single Op-amp.

Part VII Further Applications

i. Phase Shift Circuit: A circuit that is designed to output the same signal that is the input, but with a phase offset. (Where it could be seen in real world********).

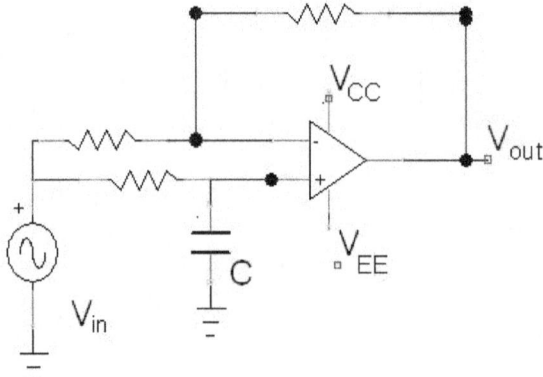

Figure 5. 11: The Phase Shifter Circuit. Output is non-inverting and non-amplifying.
http://www2.ece.ohio-state.edu/ece209/lab1_intro/lab1_intro_phase_shifter.pdf

ii. Summer-Adder Circuit: A circuit that is designed to take multiple inputs and output the summation of the inputs. (Where it could be seen in real world).

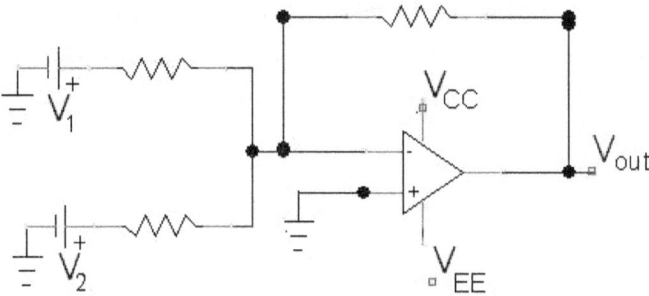

Figure 5. 12: The Summing Inverter circuit. Notice that there can be more than just V1 and V2 as inputs.
http://www.electronics-tutorials.ws/opamp/opamp_4.html

http://masteringelectronicsdesign.com/how-to-derive-the-summing-amplifier-transfer-function/

iii. The Subtractor Circuit: A circuit that is designed to take two inputs and produce as an output the difference between them. If you want to subtract the sum of two waveforms from another, you must first build an Summing circuit as seen in Figure 5. 12. (Where it could be seen in real world).

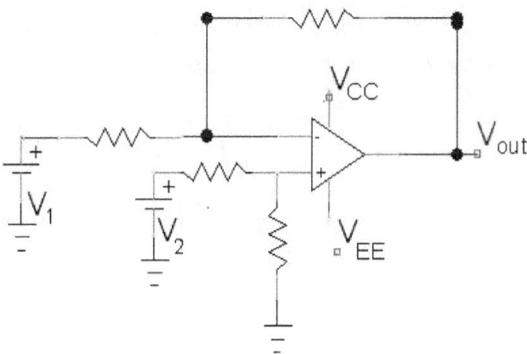

Figure 5. 13: The Subtractor Circuit. Also called a differencer, this circuit will only take two inputs.
http://www.electronics-tutorials.ws/opamp/opamp_5.html

iv. Diode and Zener Diode Tester: These circuits allow you to test if a diode is still within working parameters.

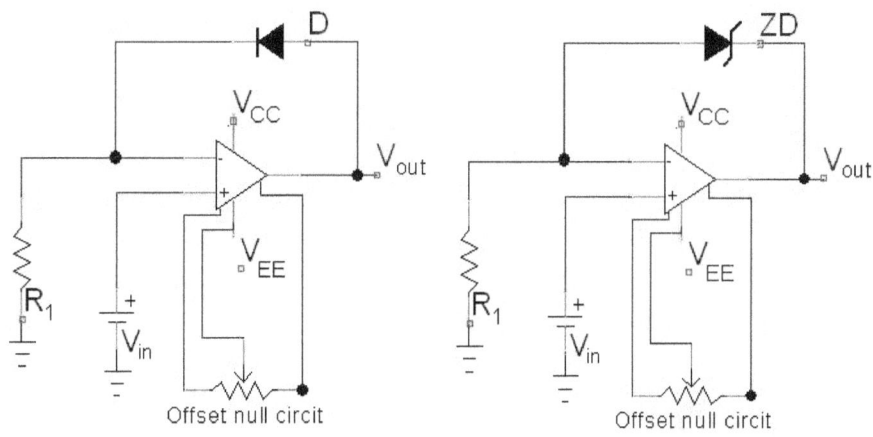

Figure 5. 14: Diode Tester circuit (left) and a Zener Diode Tester circuit (right).
http://www.wiringcircuit.com/meter/Zener_Tester_1548.html

v. Photodiode Detector: Used to sense when light is incident on a photodiode.

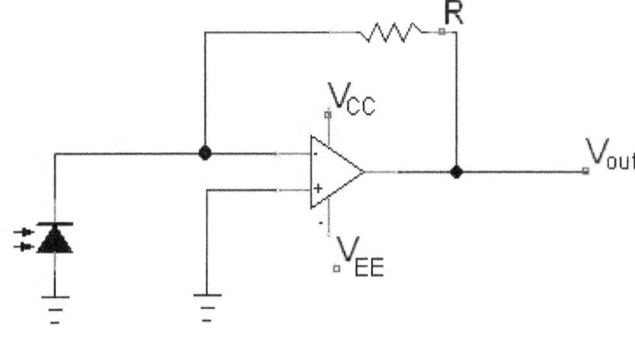

Figure 5. 15: Photodiode Detector.
http://www.electroschematics.com/6239/invisible-alarm/

vi. Window Detector: This circuit will swing the output to saturation when the input signal has a value within some minimum and maximum value.

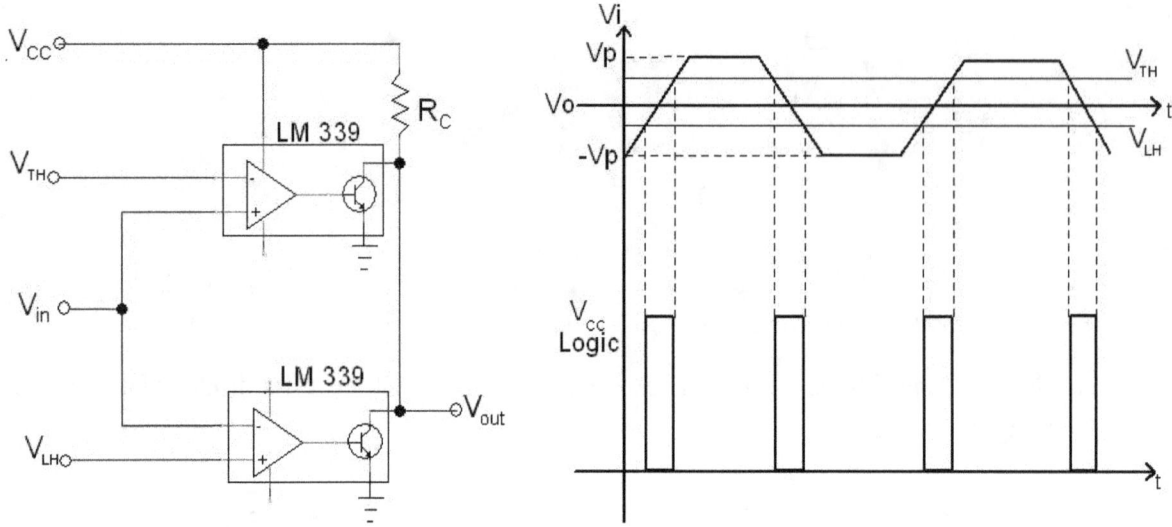

Figure 5. 16: Window Detector circuit (left) and its input compared to the output (right).

vii. Precision Full and Half wave rectifier: Unlike the diode version where the voltage drop to activate the diode caused gaps in the output signal, these circuits will not.

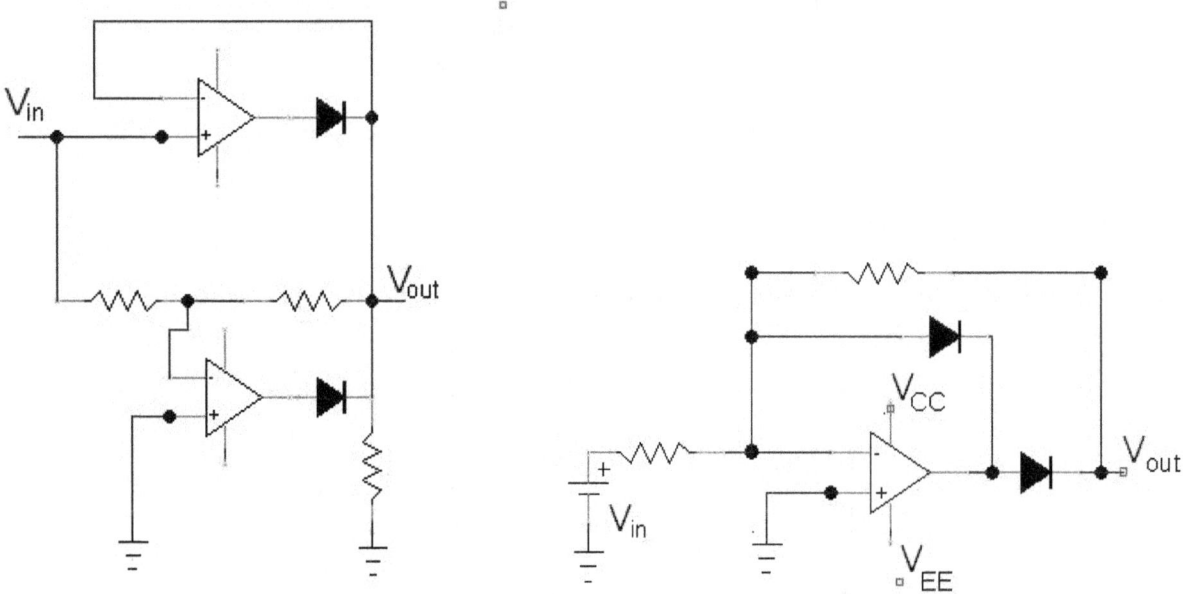

Figure 5. 17: Precision Full Wave Rectifier (left). Precision Half Wave Rectifier (right).

Part VIII Prelab Questions

(1) In Part III, calculate the answer for the first step, and fill in the box with your answer. Keep in mind what are typical values of resistors when you write your answer.

(2) Derive the Equations (5. 13) and (5. 14) and fill them in the space provided.

(3) In Part V, work through the first 4 steps of the procedure. Calculate the values asked for in each step and fill in the boxes. Keep in mind that you may have to choose slightly different values for the components based on availability when you work through the lab for real. Keep all your work in algebraic form to make it easier to make corrections on the fly.

Lab 6 - Introduction to the Arduino

Goals: Become familiar with the properties of the Arduino microprocessor system. In the process you will investigate:

1. Capabilities of the Arduino
2. Programing the Arduino
3. Control of Digital Inputs and Outputs
4. Pulse Width Modulation
5. Live Plotting Data in Excel

List of Equipment and Parts

1. Arduino and USB Cable
2. Function Generator
3. Oscilloscope
4. Computer
5. 2 Resistors (200Ω, 300Ω)
6. 1 LED
7. 1 Potentiometer (10kΩ)
8. PLX-DAQ software

Introduction and Overview

The Arduino is an open-source hardware platform designed to connect a computer to the physical world. It is designed to operate either independently or with connection to a computer. It can sample and respond to a variety of inputs such as switches and sensors and can control a variety of outputs such as lights, motors, and other actuators. It is designed around a simple programming environment (AVR C++) similar to C++ that enables users to quickly write programs (sketches) and make use of libraries to utilize its rich functionality. This enables the Arduino to communicate with a PC and to interact with a variety of hardware boards (shields) and external devices. The open-source nature of the Arduino hardware and software enables the user to quickly and cheaply develop applications.

Arduino Hardware Overview

The Arduino comes in several forms according to size, capability, and cost. We will use the Arduino Mega 2560 that is shown in figure 1. The Arduino Mega 2560 is based on the AT mega2560 microprocessor and has 54 digital input/output pins, 14 of which can be used as analog (PWM) outputs. There are an additional 16 analog inputs with 10-bit resolution (0-1023). Power can be provided either by the USB port or by an external power supply.

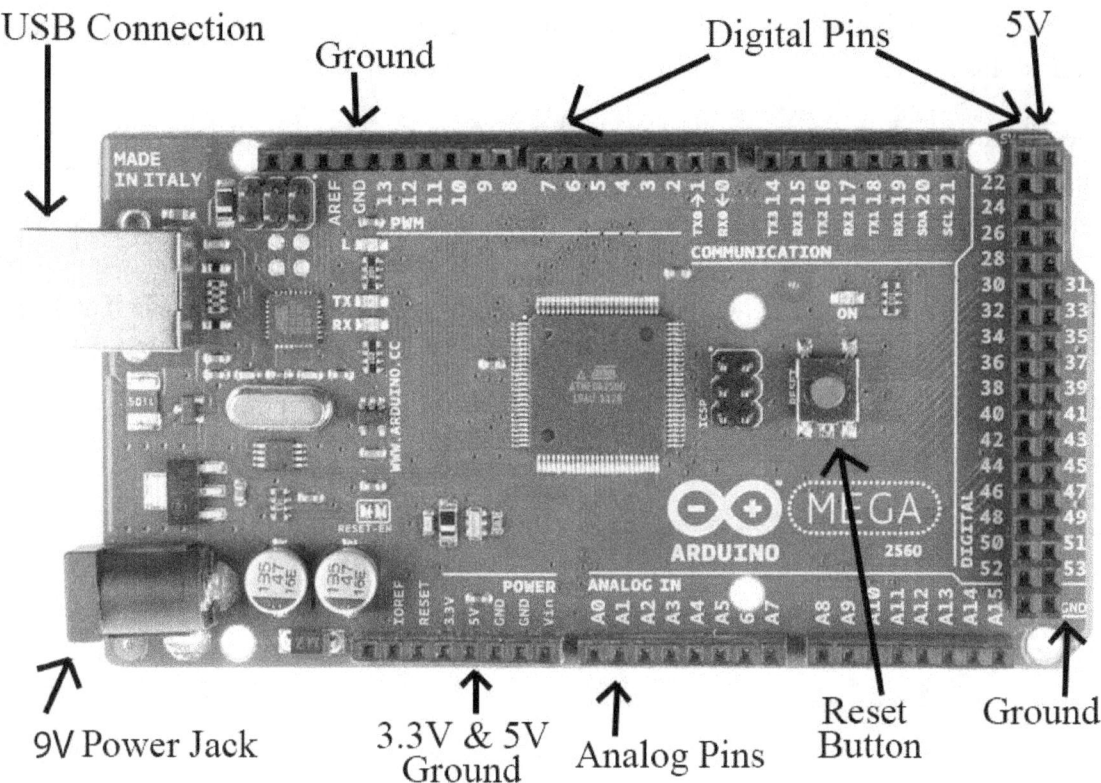

Figure 6. 1: The Arduino Mega 2560 board. It features a 16MHz processor and many more input/output pins than you can use as once.

We will be using the Arduino Mega 2560 board, as seen in Figure 6. 1. The USB plug is at the top left and the power connector is at the lower left. The CPU is located in the center of the board and various input and outputs pins are found around the periphery. Analog inputs are found along the bottom (A0-A15), Digital outputs are found along the right edge (20-53) and PWM (analog out) connections are found along the top edge (2-13). Additional pins are used for auxiliary power and communication (serial input/output).

Bare boards used as "shields" are designed so that components and pins can be soldered into place to make a separate hardware device to plug into the Arduino. However, for most prototyping and our labs we will simply use jumper wires to connect the Arduino to a miniature breadboard that will contain your circuits and various input and output devices.

Arduino Software Overview: Anatomy of a Sketch

The Arduino software consists of a compiler to turn your software (sketch) into code the microcontroller can understand and a set of libraries, which contain a set of functions that are somewhat universal. For example, having a computer interface library saves you the effort of writing math functions for each bit of code you write or having to rewrite the same computer interface code whenever you want to download a program to the Arduino. These libraries are provided with the Arduino but developers may also write a library that contains standard functions for interfacing and controlling a particular type of hardware device like a stepper motor or a WIFI transmitter.

Double-clicking on the Arduino icon on the desktop of your computer will open the interface as seen in Figure 6. 2. The pull-down menus are pretty self-explanatory and allow you to find and load an existing sketch or start editing a new one from scratch, and save your progress. There are also standard editing features like "cut", "delete", and "paste." Shortcuts with the common tasks are shown on the second line. When you finished editing your sketch you can save it and compile it using either "Sketch: verify/compile" and then clicking on the right-arrow to upload your compiled code to your Arduino over the USB cable.

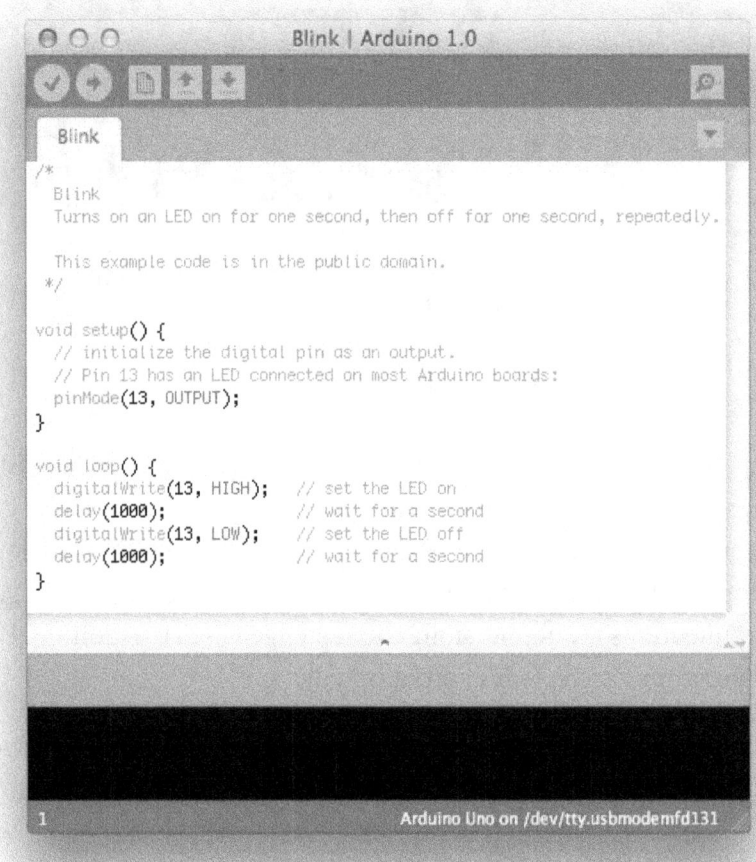

Figure 6. 2: Arduino Graphical User Interface with the sketch for blinking an LED loaded.

The Arduino interface following the loading of the first sketch we will use (blink) is shown in Figure 6. 2. The file interface and editing menus are shown along the top and the file is open for editing in the main window. Note that the color of text changes to reflect syntax. Orange are system functions, blue are system states, and black is user-defined code.

There are also various tools available such as "tools: serial monitor" which open up an old-school "serial modem" to listen for and display any output from the Arduino. We will learn about these additional features as needed. For now the steps will be pretty simple:

1) Open up and modify an existing software sketch,
2) Compile and verify the syntax of the sketch,
3) Download the sketch to Arduino and run it,
4) Save your sketch into a new folder for your lab group.

The last step is important since we don't want to overwrite an existing (working) sketch with a new, modified (broken?) version. Let's get started by loading a working sketch and modifying it.

Procedure

Part I: Getting Started by Blinking an LED

Let's begin by building a simple circuit for an LED. The Arduino will serve as the 5V power supply.

- Build a voltage divider on your breadboard to lower the voltage into the LED to 3V and plug in a LED. Attach the positive lead of your circuit to pin #13 on the Arduino and attach the negative (return) lead to the pin labeled GND. This will complete the circuit and allow the Arduino to control the output voltage.

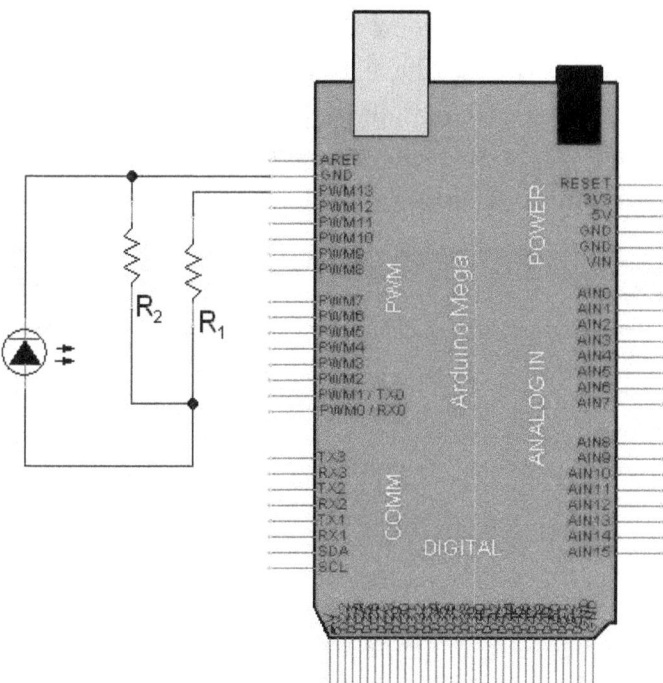

Figure 6. 3: The PWM pin 13 feeds 5V into the voltage divider when the pin is activated. The voltage divider lowers the voltage to about 3V.

- Plug in the USB power cord for the Arduino.
- Open the Arduino GUI. Before we can upload sketches to the Arduino, you must tell the software which Arduino we are using. This will only need to be done once.
  - Select **Tools: Board: Arduino Mega 2560 or Mega ADK**

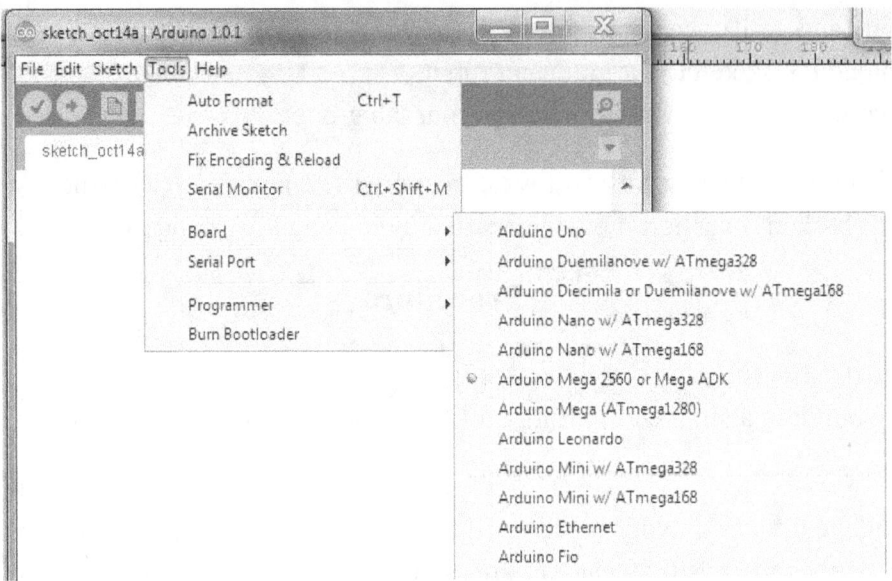

Figure 6. 4: Setting the correct Arduino board on the software GUI.

- Next, you will need to make sure the software has detected the Arduino and is currently connected to the Port that the Arduino uses. Even if the software detects the correct Port, it does not always select it, so you may have to do this again in the future. The exact number of the COM Port can vary for each Arduino depending on how many ports already exist on the computer. Warning! These steps to not apply to Linux and Mac.
  - Select **Tools: Serial Port: COMXX**

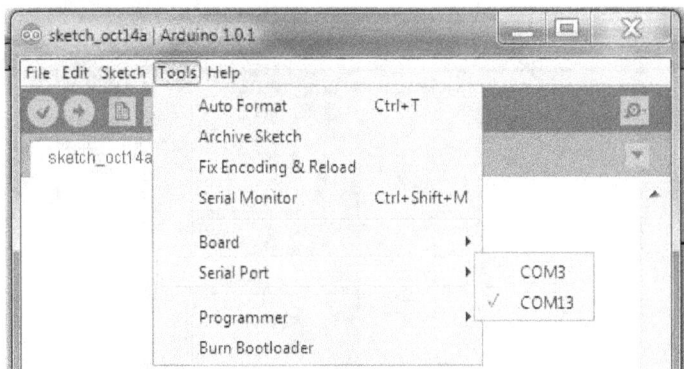

Figure 6. 5: Selecting the correct COM port of the Arduino.

- Now open the Arduino interface and select: **File: Examples: 01.Basics: Blink** to open the sketch. You should see a second GUI window open with the sketch "Blink" preloaded.

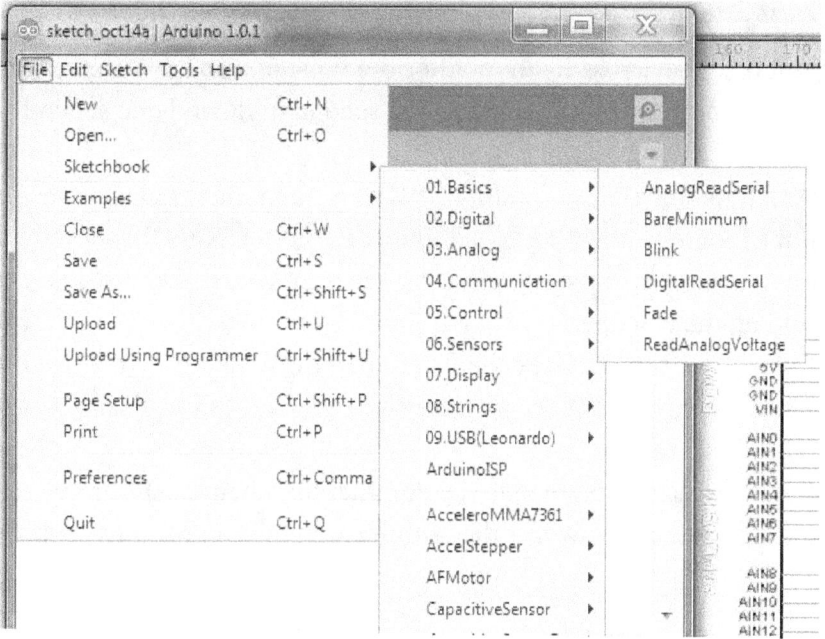

Figure 6. 6: The path to finding the sketch "Blink".

- To upload the sketch to the Arduino, you will first compile the sketch by clicking the checkmark button at the top left. Once it compiles, upload the sketch to the Arduino. Note the LED on your Arduino should flash rapidly during this process, and you should see the green bar at the lower right side of the GUI indicate the progress.
    - Note that it is technically not necessary to compile the sketch first before clicking the button to upload it, because the upload will also compile the sketch. HOWEVER, I have noticed that the compiling the sketch first decreases the chance that the software will lose its connection to the Arduino during the upload. If the upload stalls, you will have to close the GUI completely, unplug the Arduino and start over (after saving of course!).
- Although the LED is fascinating to watch, let's examine the code to see how it works. First identify the comments at the beginning enclosed by "/*" and "*/" on separate lines. All the lines of text within these symbols are considered as comments and are not compiled. Individual lines of comments are designated as "//". Note that the next line is a declaration of the variable "led" being an "int" (integer) and assigning it a value of 13. Next follows the "setup" command that initializes the CPU, then comes a pin assignment command ("pinMode") which sets pin 13 (identified by led) as an "output". The next command ("void loop") defines a loop between "{" and "}" that continues to be executed until the Arduino is powered down or we download another sketch. Its contents are:
    1) "digitalWrite" (write the value HIGH (+5V) to pin defined by variable led),
    2) "delay" (wait 1000 msec),
    3) "digitalWrite" (write the value LOW (0.0V) to pin defined by variable led),
    4) "delay" (wait 1000 msec)
    5) loop back to step 1 and repeat.

Part II: Digital and Analog Input/Output

- Now place your o-scope across R_2 to monitor the voltage through the LED. You should see a square wave with a period of two seconds: one second high, and one second low. Record the waveform.
- Modify the sketch to blink faster. Within the "delay" commands change the values in both places within the loop to something like 200 msec ("200"). Note that nothing will happen until you compile and upload the sketch using the right-arrow. Do that and describe what you see visually and on the o-scope.
- Try different values for the "HIGH delay" and the "LOW delay". Now use the "cut and paste" features of the editor to make an SOS signal: 3-dots (short flash), 3-dashes (long flash), and 3-dots (short flash).
- Finally, eliminate the delay commands (by commenting them out) from your sketch and investigate with the o-scope how fast the Arduino can turn on/off pins with the digitalWrite command.

- When you have the sketch working, edit the comment lines to reflect the purpose of your new sketch and save it in the "My Documents" folder with a new name that reflects what it does. Be sure to save this and your other sketches to a flash drive to share with you lab partners. Note that we have examined the primary features of +5V digital signals (aka TTL). These are simply amplitude (always +5V) and frequency.

Now lets examine how the Arduino processes analog signals. Computers don't really like analog signals, so they have to fake them. Consider output first. Microcomputers are so fast that they can toggle the +5V on and off with a delay (at 0V) such that the average signal level is less than +5V. This is called pulse width modulation (PWM). Let's examine PWM by re-defining pin #13 to be an analog (PWM) output. This is done with the "analogWrite" command. This command takes an 8-bit number (0-255) and maps it from 0V to +5V on the output pin via PWM.

- Begin by loading the "fade" sketch by selecting: **File: Examples: 01.Basics: Fade.** Examine the sketch. Note that each time the loop is run the brightness is incremented until the brightness reaches maximum (255) and then the "fadeamount" changes sign so that the brightness is now decremented. This reverses again and the diming/brightening continues.
- Upload the sketch and note what happens to both the brightness of the led and the output of the o-scope. Thus, an analog device can be controlled via PWM. Notice that the amplitude does not change, only the period of time that the wave is not zero. This can be used to control any motor's speed.

Part III: Analog to Digital Conversion (ADC)

In this example, we will use the ADCs on the Arduino to digitize an analog signal. The 10-bit ADCs on the Arduino allow any analog signal (0 – 5V) to be represented as a digital number (0 – 1023).

- Begin by connecting the outer pin on a potentiometer to the +5V output on the Arduino. Connect the other outer pin of the potentiometer to GND. Finally, connect the middle pin to analog pin A0. You may need to use jumpers to the breadboard to allow for connections to the potentiometer.

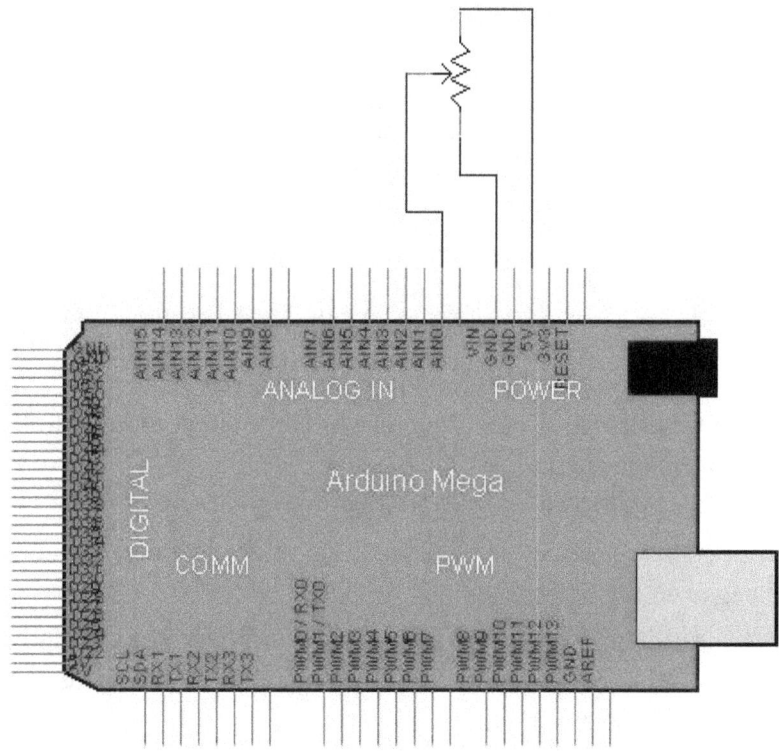

Figure 6. 7: A potentiometer feeds voltage to the analog pin. The analog pin samples the voltage from the 5V out.

- Now load the sketch by selecting: **File: Examples: 01.Basics: AnalogReadSerial.** Examine the sketch. Note the command "Serial.Begin" which sets the baud rate of the serial output to the USB port. This command also starts a process on the host computer that reads the serial output from the Arduino via the "analogRead" command.
- The "Serial.println" command prints the values that are read to the serial monitor. To access the serial monitor window, go to **Tools: Serial Monitor** or hold "Ctrl Shift M".
- Connect the DMM across the potentiometer so that you can monitor the voltage drop across the pot.
- Compile and upload the sketch. Change the potentiometer and see how the values change in the serial monitor. At this point, you could copy and paste the values you have read to an excel sheet for analysis, but we have a better way.

- Close the Serial Monitor, and open the program PLX-DAQ. It should start Excel and warn you that you are about to initialize ActiveX. Click "OK." You should see a window shown in Figure 6. 8.

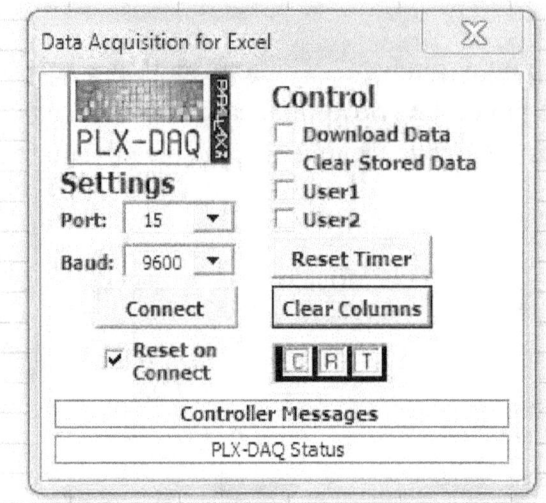

Figure 6. 8: The PLX-DAQ GUI. Baud should always be set to 9600, but the port you will have to set yourself.

- Set the Baud to 9600. Set Port to the COM number for your Arduino.
- We have to make a few changes to the sketch to pass the data correctly to this program. Before void setup(){ add these lines:
    - int row=0; //defines row as an integer.
    - float sec=0.0; //defines sec as a floating point (decimal) value.
- Beneath Serial.begin(9600); add these lines:
    - Serial.println("CLEARDATA"); //Tells PLXDAQ to clear all previous data.
    - Serial.println("LABEL,Time,Sec,sensorValue"); //Sends column headings.
- Also, replace Serial.println(sensorValue); with
    - Serial.print("DATA,"); //Tells PLXDAQ that what follows is data.
    - Serial.print("TIME,"); //Sends the time to PLXDAQ.
    - sec=millis()/1000.0; //Calculates time in seconds from start of program.
    - Serial.print(sec); //Sends time in seconds from start of program.
    - Serial.print(","); //Sends a spacer between sent values.
    - Serial.println(sensorValue); //Sends final value (sensor value) to PLXDAQ.
    - row++; //Moves input to next line of Excel sheet
- Compile and upload the sketch to the Arduino. Then, in the PLX-DAQ program, click "connect." You should now see values populating columns in Excel.
- Insert a scatter plot in excel, plotting the analog voltage values vs. time in seconds. When you turn the potentiometer, you should see a response on your graph in real time.

Part IV: Bonus! Investigating the Smoothness of the Potentiometer

Many students have noticed that the potentiometers seem to have places in their rotation (especially near the ends) where the potentiometer behaves in much more abrupt fashion, jumping from one voltage to another without smoothly transitioning between them. You will now investigate just how continuous your potentiometer is.

- On your Excel sheet, now that you have it plotting in real time the voltage from the potentiometer, add two new columns: the derivative of volts and the double derivative of volts. These will correspond to the velocity of turning of the pot and the acceleration.
- Next, as smoothly as you can, turn the potentiometer from one extreme to the other and examine the plots. Look for sharp spikes in acceleration that would indicate discontinuous motion. Do these seem to happen near the ends of rotation?
- Add a new column that makes a running average of volt readings to smooth your data. Experiment with how many samples to average together so that you can still clearly see spikes in your data, but the noise is removed.

Part V: Bonus! Bonus! Make a Dimmer from the Potentiometer

- Combine Parts II and III (the fader and the potentiometer) to make a dimmer switch. Make the LED brightness in the fading sketch dependent on the amount of voltage coming from the potentiometer. Give it a maximum brightness at maximum voltage and vice versa.

Part VI: Summary

In this lab, you have examined some of the properties of the Arduino. You have examined three different sketches that access the three types of input/output available on the Arduino. You have also learned how to track data in real time from the Arduino using PLX-DAQ in Excel.

Lab 7 - Arduino Output Control

Goals: Become familiar with several possible outputs that the Arduino can control:

1. 7-Segment Display
2. SD Card Write
3. Motor Control
4. Relay

List of Equipment and Parts

1. Arduino and Cables
2. Computer
3. Oscilloscope
4. DC Power Supply
5. 1 Resistor (100Ω)
6. Jumper Wires
7. 7-Segment Display
8. Motor Control Shield
9. SD Card Shield and SD card
10. Relay Module
11. Small Speaker
12. Stepper Motor
13. DC Motor
14. Case Fan

Introduction and Overview

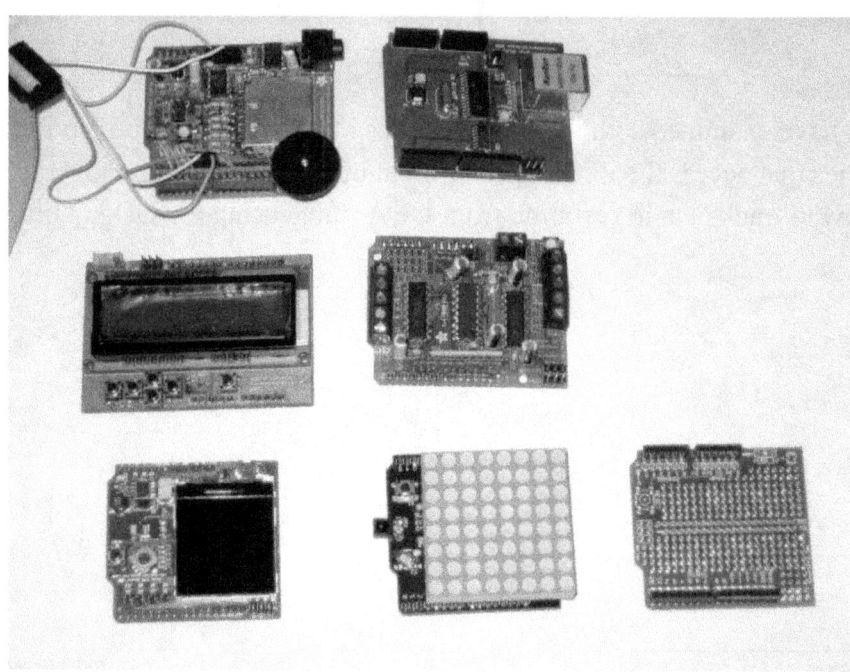

Figure 7. 1: Arduino Shields. Top - Wave, Ethernet. Middle - LCD Key Pad, Motor Control. Bottom - Color LCD and Joystick, Custom LED Matrix, and a Prototype shield.

The Arduino is capable of controlling many devices, more than we have in this lab. Some time spent on the internet can discover Arduino controlled parking sensors, security cameras, robots, and many more. This flexibility gives the Arduino much of its value.

To cut down on some of the complexity, there are so-called shields that plug into the Arduino. Think of a shield as breadboard with a permanent circuit designed to a specific use. There are a few dozen shields, with more being made each month.

Procedure

Part I: 7-Segment Display

In the previous lab, you learned to control one LED; not very useful for displays. The most common form of display, until LCDs took over, is the 7-segment display. It gets its name from the seven LED segments that allow the user to display numbers. There can also be a decimal point or two in either the bottom right or left corners. There are several configurations, but the version we will be using is in Figure 7. 2.

Figure 7. 2: The 7-Segment Display we will be using with corresponding pinout description and segments labeled.

- Start the Arduino software and find the sketch "_7_Segment Count Down". Notice how each number is created by referring to an array at the beginning of the sketch. The sketch will make a countdown, running through each number for about a second.
- This will only work, however, if the appropriate pins are connected on the Arduino to the display. We can see from the sketch that the pins 2 – 9 are used. Place the display straddling the middle of a breadboard and connect the following Arduino pins to the corresponding contacts on the 7-segment display (Table 7. 1).
- We will need to run both ground pins through a resistor to keep from overloading the LEDs. Consult
- Table 7. 2 and Table 7. 3 to determine what value of resister you will need and place the resistor in series with the ground pins.

Arduino Pin	7 Segment Pin Connection
2	7 (A)
3	6 (B)
4	4 (C)
5	2 (D)
6	1 (E)
7	9 (F)
8	10 (G)
9	5 (DP)

Table 7. 1: Table of Arduino pins and corresponding pins on a 7-segment display.

Electrical / Optical Characteristics at TA=25°C

Symbol	Parameter	Device	Typ.	Max.	Units	Test Conditions
λ_{peak}	Peak Wavelength	High Efficiency Red	627		nm	I_F=20mA
λ_D [1]	Dominant Wavelength	High Efficiency Red	625		nm	I_F=20mA
$\Delta\lambda_{1/2}$	Spectral Line Half-width	High Efficiency Red	45		nm	I_F=20mA
C	Capacitance	High Efficiency Red	15		pF	V_F=0V;f=1MHz
V_F [2]	Forward Voltage	High Efficiency Red	2.0	2.5	V	I_F=20mA
I_R	Reverse Current	High Efficiency Red		10	uA	V_R=5V

Table 7. 2: Characteristics of the 7-Segment Display taken from the Spec Sheet.

Absolute Maximum Ratings at TA=25°C

Parameter	High Efficiency Red	Units
Power dissipation	75	mW
DC Forward Current	30	mA
Peak Forward Current [1]	160	mA
Reverse Voltage	5	V
Operating / Storage Temperature	-40°C To +85°C	
Lead Solder Temperature[2]	260°C For 3-5 Seconds	

Table 7. 3: More Characteristics of the 7-Segment Display taken from the Spec Sheet.

> ➤ Once the circuit is built, compile and upload the sketch. Hopefully, the display is now counting down from 9. Notice that you can restart the countdown by pressing the reset button on the Arduino board itself.
> ➤ Change the sketch to make the countdown proceed at twice the speed. Then change it to countdown at half the speed.
> ➤ Now change the sketch to make it count up to 9 instead of counting down.
> ➤ Finally, modify the sketch to display three non-sequential numbers with a one second delay between each number.

Part II: Motor Control

Now we can try our first shield; the Motor Control Shield. This shield allows for the control of standard DC motors as well as Stepper motors. This model is limited to 600mA maximum constant current draw with a 1.2A maximum short term peak current draw. Newer models exist through Adafruit that can handle at least double the current. Connect the shield as shown Figure 7. 3.

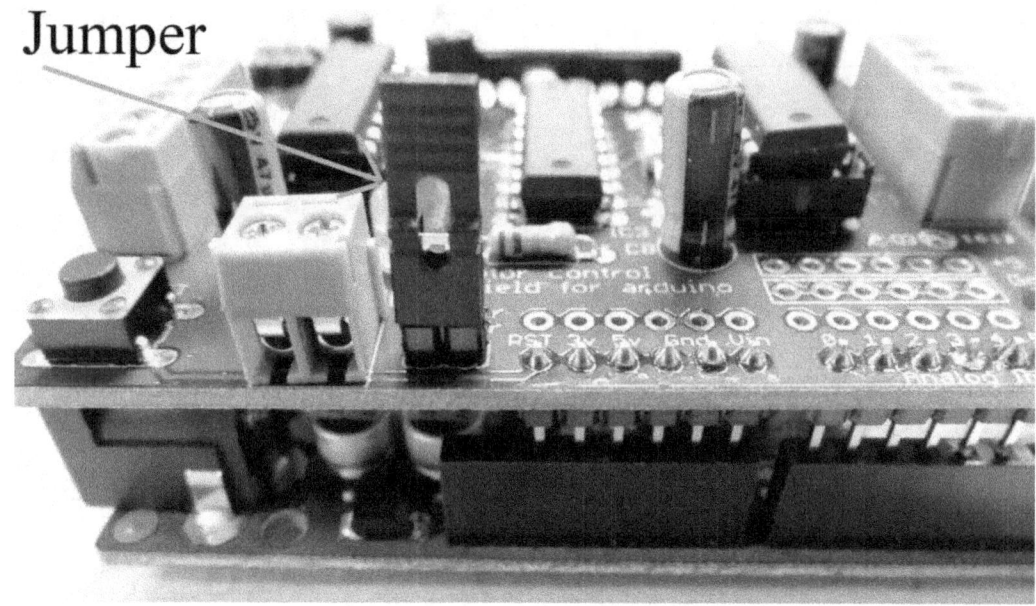

Figure 7. 3: Connecting the Motor Control Shield to the Arduino.

- ➢ Load the sketch **Examples: AFMotor: MotorTest** and plug the DC motor (Figure 7. 4) into the M4 port. The DC motor can run in either direction at high RPM, but has little torque.
- ➢ Run the sketch and you should see the motor run forward and then backward.
- ➢ Change the sketch so that it runs the motor out of M2, connect the motor there, and run it.
- ➢ While the motor is running, remove the black jumper featured in Figure 7. 3 (rectangular piece just to the left of center). This is a convenient on/off switch if you don't want your motor running constantly.

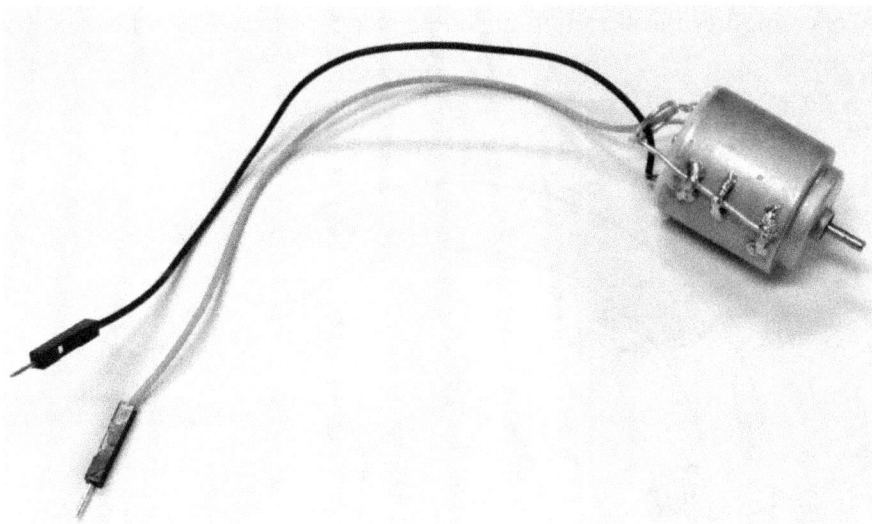

Figure 7. 4: The DC motor. Capable of high rpm but has low torque.

- Modify the sketch to make the motor run faster by changing the motor speed in the sketch. If you keep increasing that number until the motor draws more current than the Arduino will allow. How does the motor respond?
- Finally, give the DC motor power on the lowest setting and see if you can turn the motor axel in reverse with your fingers. Notice that the motor has very little torque.

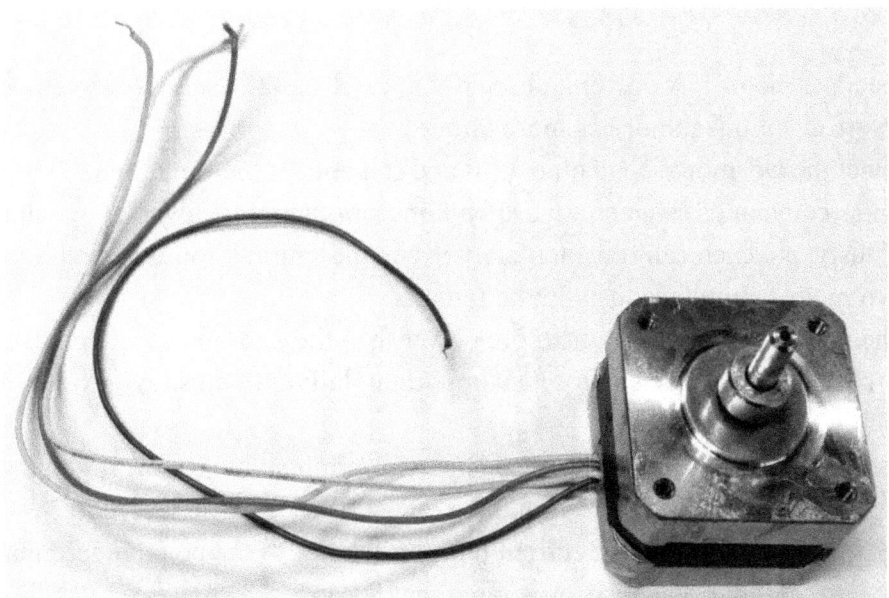

Figure 7. 5: The Stepper Motor. Only very low rpm, but high torque. Useful for moving a precise motion.

- Next, retrieve a stepper motor (Figure 7. 5) and connect the wires in the order of Red, Yellow, Skip Ground, Green, Brown across two M terminals. Open the **Examples: AFMotor: StepperTest** sketch, and run the stepper motor.

- Once that works, modify the sketch to make the stepper run twice as fast and complete on full circle.

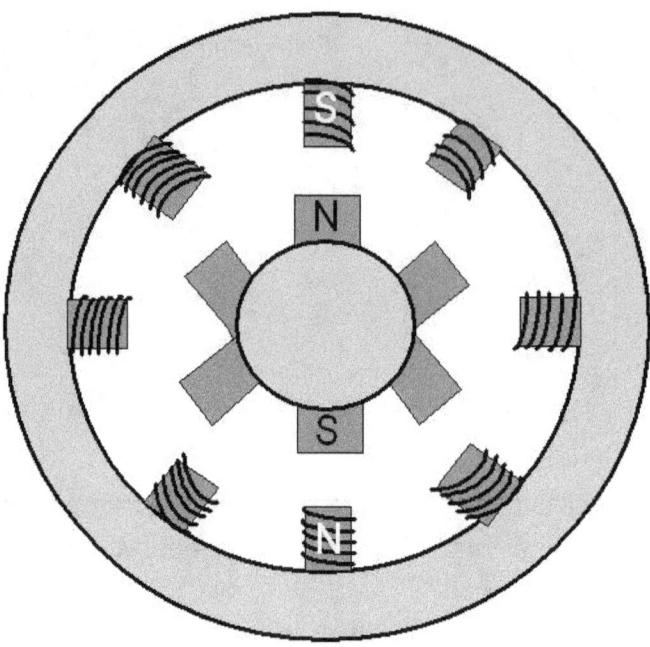

Figure 7. 6: Internal Diagram of the Stepper Motor. The top and bottom coils have been powered to create magnetic fields to attract the correct magnetic teeth. When the coils are activate in sequence, the motor rotates

- Give the stepper motor low power and see if you can turn the axel in reverse with your fingers. Notice that this motor has more torque.
- Now connect the DC motor on another port and run both at the same time. Notice that the Arduino runs commands sequentially, so both motors cannot be given commands simultaneously, but each can run alternately. Keep these limitations in mind if you want to control two motors simultaneously in the future.
- Finally, the motor shield can handle more power than the Arduino can. Remove the jumper pictured in Figure 7. 6. This is *extremely* important; failure to do so will result in damage to the Arduino.
- Immediately to the left of the jumper is a blue two-terminal connector. If you look close, you can see that one is marked with a '+' and the other with a '-'. Connect the 12V from the power supply through a 100 ohm current limiting resistor to the positive terminal and ground to the negative terminal. The resistor is important because the shield can handle only 600mA of current at 12V.
- Try the DC and stepper motors again and observe the difference now that they are powered by a 12V supply.

Part III: Relay Control

A relay is used when keeping the power supply separate from the controller is paramount. For our use, we will be using the 5V ~18A line from the power supply rather than the 5V pin from the Arduino to drive a motor. This way, we should be able to exceed the power limitations of the Arduino, while protecting the Arduino from the high current of the power supply.

➢ Remove any shields connected to the Arduino, and retrieve the 4-channel relay (Figure 7. 7).

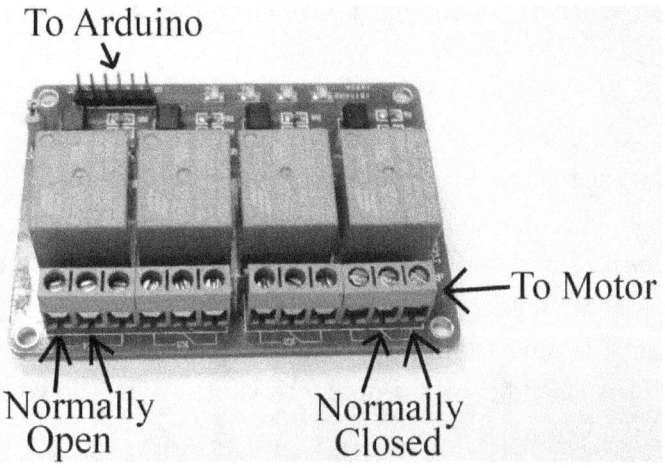

Figure 7. 7: The 4-Channel Relay Module. Each output has a normally open and a normally closed. A Normally open would break the circuit until the Arduino activates the relay and vice versa for the normally closed.

➢ You will need to connect the Arduino to the module via male and female jumper wires. Connect the 5V from the Arduino to the Vcc, the Ground to the Ground, and digital pin of your choosing on the Arduino to IN1. We can now control the switch at K1 with a HIGH or LOW signal from the pin into IN1.
➢ Plug the DC motor into K1 so that it is in the Normally Open side of the switch and connect the 3.3V line from the power supply as shown in Figure 7. 8. For R1, choose a few ½ watt resistors in a range from 2 to 10 Ohms. If you are going to greatly exceed ½ watt for a long period, you might switch up to a 1 or 2 watt resistor

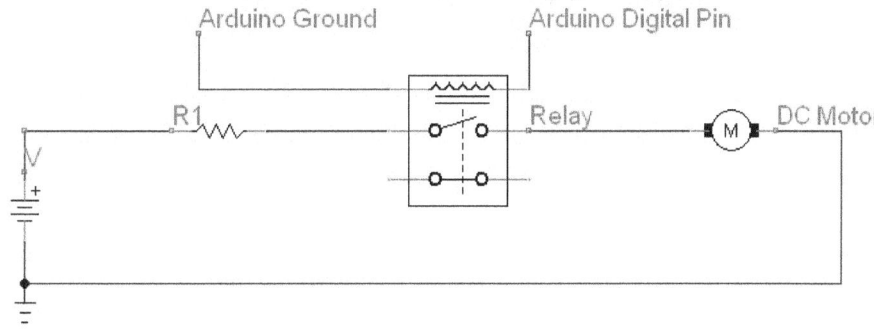

Figure 7. 8: The circuit diagram for connecting the Relay to the Arduino and a DC motor.

- Open a sketch you used for blinking an LED, and modify it to send the signal to the pin you chose. The relay is switched ON when the pin is in the LOW setting, so change the time on the LOW setting to about 10 seconds. When the relay switch is turned on, you should hear an audible click/pop sound and a red LED should light to indicate that it worked (and, hopefully, your motor should start running).
- This DC Motor should reach maximum efficiency at ~1.3 Amps. Do your trials seem to confirm this? The motor is rated to just below 5V, but I have tried it out at 5V and, for short periods of time, the motor can handle it. Try running the motor at 5V and see if it makes a difference in speed.
- Now, remove the DC motor and connect one of the computer case fans Figure 7. 9 we have in its place. If the fan has only two wires, they are self-explanatory. If there are more than two, look for the red and black wires as the power. The other wires are a tachometer and/or a PWM control wire. These fans are 'brushless' motor, meaning they have a permanent magnet at its core, and they are quite robust. Do not worry about supplying too much current to these fans, they can take it.
- Most fans are 12V fans, so use the 12V line (however, some are rated for lower voltage so read the fan first!), and try out different resistors as you did with the DC motor. See how fast you can get the fan to go.

Figure 7. 9: Case fan from a computer. Red and black wires are the power and ground.

Part IV: SD Card

Should you want to data log on the Arduino without being tethered to your computer, the SD Card Shield is the solution.

Figure 7. 10: The SD card reader shield and the SD card.

- Find the SD card shield shown in Figure 7. 10 and connect it to the Arduino in the same manner as you connected the Motor Control shield. Note that the card comes with a battery to supply power to the board for reading and writing to the SD card, but the Arduino will still need to be powered from either a USB or a 5V line.
- Because we are using the Arduino Mega, we will need to make a small change to one of the library files before the shield will work properly.
    - Navigate under the Arduino folder to **libraries: SD: Utility** and find the **Sd2Card.H** file.
    - Open it in a text editor and find the line that reads **#define "MEGA_SOFT_SPI 0"** about a fourth of the way down.
    - Change the 0 to a 1 and save the file. This will have to be done on any computer you wish to run the SD card shield.
- Next, start the Arduino GUI, and load the sketch **SD: CardInfo**. Two changes must be made to this sketch before the shield will run properly with the Mega.
    - Find the chipSelect and pinMode sections of the sketch, and change the numbers to those listed in the comments that correspond to a Mega and Adafruit shield.
    - Finally, insert your SD card into the shield and run the sketch. Information about the card will be displayed in your serial monitor.
- Once you know the shield is working properly, load the sketch **ReadWrite** and make the same changes to pinMode that you did on the previous sketch. chipSelect will have to be changed to 10, but it is not obvious where.
    - Look for if (!SD.begin(4)) {

- o Change the 4 to a 10
- ➢ Run the sketch and open the serial monitor to see something like Figure 7. 11. Now that you can read and write to the SD card, change what you are writing to the card. Write the digits, 1 to 10 and modify the sketch to give a heading before displaying what was read from the card.

Any text can easily be stored and retrieved. In theory, the SD card can store pictures, video, etc…, but it takes a bit more programming than we are going to get into to retrieve the pictures.

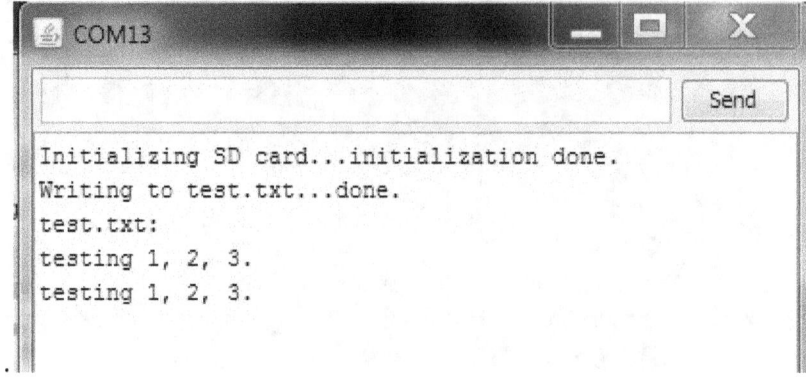

Figure 7. 11: The output on the serial monitor of the sketch, ReadWrite.

- ➢ Try the other sketches under the SD folder (remember to change the chipSelect and/or pinMode on each). Start with the DataLogger, which will create a datalog.txt on your card, recording readings from three sensors (which are not installed, so just dummy data).
- ➢ The sketch, **Files,** will allow you to search for a specific file on the card.
- ➢ **Listfiles** will give a list of what is on the SD card.
- ➢ **FileDump** can open a file that is on the SD card such as a text file
- ➢ If you want to remove a file from the card, use this code:
  - o // delete the file:
  - o Serial.println("Removing example.txt...");
  - o SD.remove("example.txt");

Part V: Tones and Speakers

Suppose you want to give an audible feedback to a user, you might need to make sound. For that, we have piezo buzzers and several speakers pulled from old computers.

Figure 7. 12: A Speaker!

- Open the sketch **Digital: ToneMelody** and connect the piezo buzzer to pin 8 and ground.
- Run the sketch and you should hear a melody played once from the buzzer.
- Switch to the speaker by connecting the red wire to pin 8 and black to ground and play the melody again.
- Modify the sketch to make a new melody.
- Once done, there are many other sketches to make tones. Try them out!

Part VI: Bonus! Potentiometer Motor Control

- Start with the section in Lab 6 where you used a potentiometer to feed in a variable voltage into the Arduino. Combine this sketch with the sketch to run the DC motor. Make the speed of the DC motor conditional on the turn of the potentiometer (all the way to one side is fast, all the way to the other side is slow/off).
- Possibly useful code: Mapping one range of values (old_x) to another range (new_x).
    - new_x=(int)map(old_x,old_min,old_max,new_min,new_max);

Part VI: Bonus! Bonus! Display the Speed of the Motor with LEDs

- Building on the last Bonus section, connect a 7-segment display in such a way so that it displays a number, 1-9, representing the speed of the motor. So, as you start slow and turn your potentiometer increasing the speed of the motor, the display will increase the number until max speed is reached at 9.

Part VII: Turkey! Processing to make a GUI to control LEDs

Processing is a JAVA based language that was developed by the same people who wrote the Arduino GUI. It includes a library to interface with the Arduino and has been out long enough to have libraries for creating easy GUI's.

- Start up Processing by clicking the shortcut and look for **MouseLEDs**. Load the sketch.
- Look for the line that says **myPort = new Serial(this, "COM14", 9600);** and change COM14 to the COM port your Arduino is using.
- Next, go the Arduino GUI and load up **MouseLEDs**. Note what pins are used in the sketch and connect LEDs to those pins (and back to ground of course). Make it look good by using Blue for low numbers, Green for middle numbers, and Red for large numbers.
- Upload the Arduino Sketch.
- Run the Processing Sketch.
- You should now see a window on your screen. As you move the mouse over it, Processing will track the mouse movements and send that data to your Arduino. The Arduino will then use that data to light up the appropriate LEDs.

While Processing is not necessary for the class, it opens up more possibilities than the Arduino alone can give.

Part VI: Summary

We have covered several types of shields and sketches. There are many more in existence with many more being developed each year. While not strictly necessary, shields provide a convenient and fast way to perform more complex tasks.

Lab 8 - A Sensor Buffet for the Arduino

Goals: Become familiar with several possible sensors that the Arduino can sample

1. Read a Temperature Sensor
2. Read a Photo-resistor (Light Sensor)
3. Read an Infrared Motion Sensor
4. Read a Tilt Sensor
5. Read an Ultra-sonic Range Finder

List of Equipment and Parts

1. Arduino and Cables
2. Computer
3. Oscilloscope
4. 2 Resistor (200Ω, 300Ω)
5. 3 LEDs
6. 1 Accelerometer (MMA7361)
7. 1 Ultrasonic Rangefinder (HC-SR04)
8. 1 Temperature Transistor (TMP36)
9. 1 Photo-Resistor
10. PIR Motion Detector

Introduction and Overview

The Arduino is capable of monitoring or sampling a wide variety of sensors. Recall that it can sample both digital and analog inputs. These sensor signals can be output to your computer's serial monitor or stored on an SD card for later use or analysis. We will take a look at several sensors in this lab but these only scratch the surface of what is available. The goal is to provide examples, which may prove useful in your semester project. In this lab, we will primarily use the serial monitor to display output. The next lab will make use of the SD card and some simple analysis using Excel.

Figure 8. 1: (left) Infrared Motion Sensor, (left center) A CdS (Cadmium –Sulfide) Photo-resistor, (middle) A 3-axis Accelerometer, (right middle) Temperature Sensor (TMP 36GZ Diode). (right) Ultrasonic Rangefinder

Procedure

Part I: Temperature Sensor

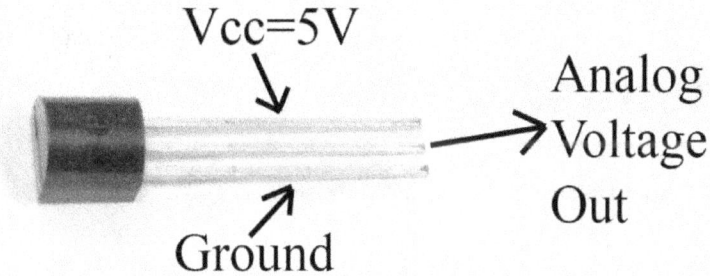

Figure 8. 2: The Temperature sensor and the pin layout.

The TMP36 temperature sensor provides a voltage output that is proportional to its temperature. Its accuracy is about +- 1 °C over a range of -40 °C – 125 °C. The constant of proportionality and the zero point vary slightly from one sensor to another but is roughly 10 mV/°C. Inside is a transistor whose collector to emitter voltage varies predicatively with temperature. The Vcc pin supplies power to both the collector and base, while the analog out pin is the emitter voltage.

- Begin by connecting the sensor to ground and to +5 VDC using the female/male jumper wires. These are the right (GND) and left (VDC) pins when viewed from the bottom. Connect the output pin in the middle to one of the analog input pins on the Arduino to measure V_{out}.
- Open the sketch **File: Sketchbook: Temp**, and open the serial terminal output on the PC. Once you have it working, we can obtain a better calibration.
- Since each sensor can vary a bit, we will need to calibrate the sensor by measuring the output voltage at two reference temperatures. Using these values, you can compute the slope (proportionality constant) and intercept (zero point).
- We have an accurate Mercury thermometer that can provide an accurate measurement of room temperature. Write down the temperature and record an average of the corresponding output voltage.
- The second temperature standard we will use will be ice water. When ice is placed into a glass of water the water will cool to 0 °C. Place the thermometer into the ice water and record the temperature.
- Next, *carefully* dip the sensor into the ice water **without** getting the electrical leads wet. We do not want to short the sensor! Record the output voltage from the serial monitor.
- Use these values to compute the calibration constants and edit the sketch to use the two new values. Restart the sketch and verify that the sensor is working properly.

Part II: Photo-resistor

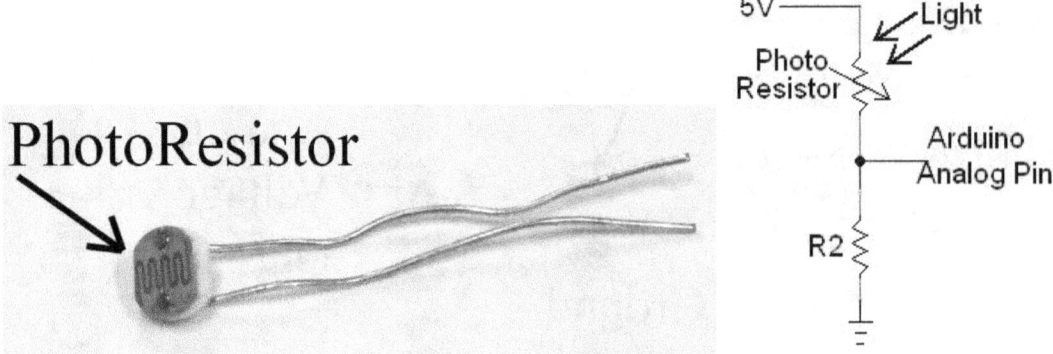

Figure 8. 3: Voltage divider circuit for the photo-resistor (aka photoconductor). Note that R2 can be chosen so that the output voltage is in the midrange of the Arduino input

Now, let's try the photo-resistor sensor. The photo-resistor is just what it sounds like. It is made from a strip of Cadmium-Sulfide, a semiconductor. As light falls on the sensor, electrons are excited into the conduction band causing the resistance of the strip to drop. That is, the resistance drops as more light falls on the sensor. The spectral response depends on the band gap but a CdS device peaks in the visible. A Cadmium-Selinide (CdSe) semiconductor has a smaller band gap and so its peak response is in the infrared.

➢ We will make a simple voltage divider circuit with the CdS sensor as one resistor and the second resistor chose so that the output will be about 2.5V, the midrange of the analog input of the Arduino.
➢ A good place to start is about 100k Ohm, depending on the light level. However, for low light level the resistor should be about 1M ohm and in sunlight a value of about 100 ohms will make the output within the range of the Arduino. Potentiometers can help here.
➢ Build the circuit shown in Figure 8. 3, and connect the output voltage to one the analog inputs. Recall that these inputs are connected to an analog to digital converter with a range of 0 – 1023.
➢ Use a simple "analog read" type sketch to input the data from the sensor to PLX-DAQ and plot the light level over time. Adjust the resistor so that the output voltage is about half scale or ~500 digital units and record this resistance as the value needed for indoor lighting.

➢ Use a flashlight to provide a much brighter source. From the graph, see if you can determine the response time of the photoresistor.

➢ With the sensor receiving the most light from the flashlight, find a new value of resistor that will set the output voltage to half scale as you did for indoor lighting. This will be the resistor you would use for outdoor use on a sunny day.

- Next, shine a laser on the sensor, and, from the graph, see if the response time is the same or different from what you measured previously.

- Alter the resistor to bring the output voltage in the mid-range as you have before and record its value for extremely bright light.

- We have three total colors of LEDs. Test the other colors on the sensor. Is it more sensitive to one of the colors over the others?

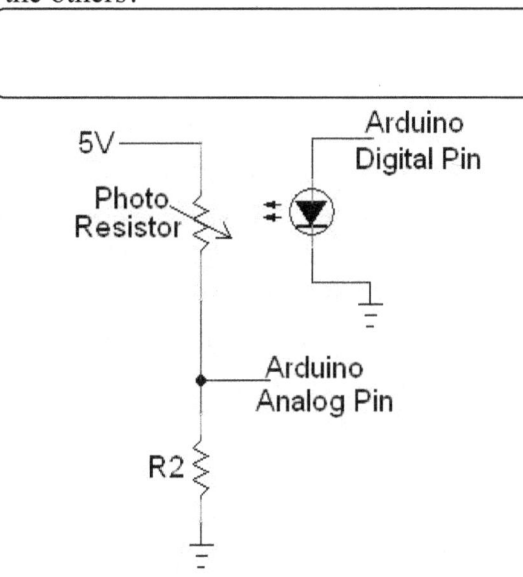

Figure 8. 4: A photoresistor is physically near an LED beneath a box to block out background light. When the LED is on, the photoresistor will change resistance.

- Finally, you will test the sensor's low light ability. Build the setup shown in Figure 8. 4. Make sure the LED has a resistor in series that will lower the brightness of the LED to a point that is hard to see in ambient light, but does actually produce a light.
- Enclose the LED and sensor in a box, to block out most of the ambient light and program the LED to stay on constantly.
- While watching the data plot in PLX-DAQ, adjust the resistor until the output voltage is set to approximately 500. Record its value for low light setting.

- Make the LED blink slowly, and examine its response time. Does it take longer to change value? What is the rate of change in low light

Part III: Infrared Motion Sensor

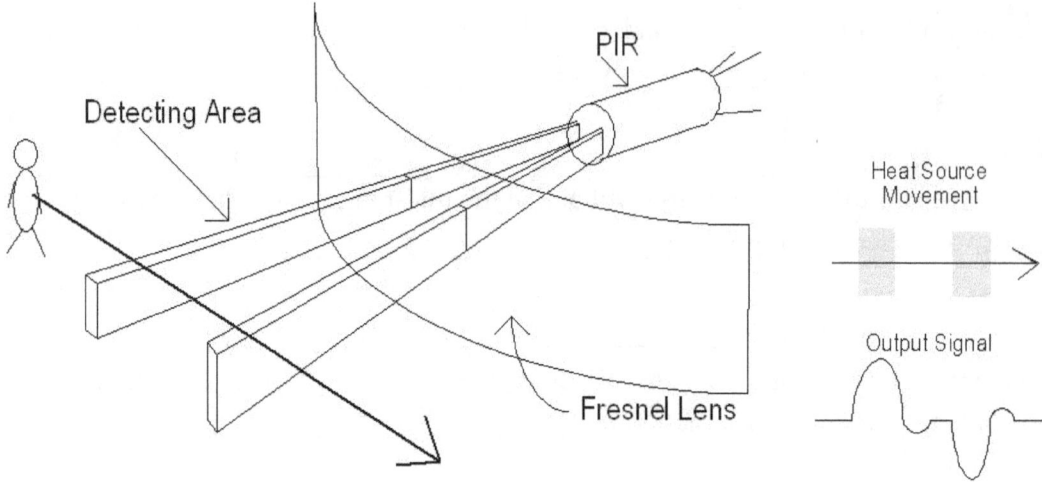

Figure 8. 5: Cartoon showing the output difference signal from the infrared motion sensor.

The infrared motion sensor consists of a "pyroelectric" material, whose bandgap is very small, and is sensitive to wavelengths of about 10 microns. The sensor is, then, sensitive to radiation at mid infrared wavelengths. Two sensors are connected to an OpAmp differencing circuit in order to sense changes in the infrared radiation shining on the sensor. Each of the two sensors sees a different direction so that as a source of infrared radiation moves across the detecting area the difference signal from the two sensors changes.

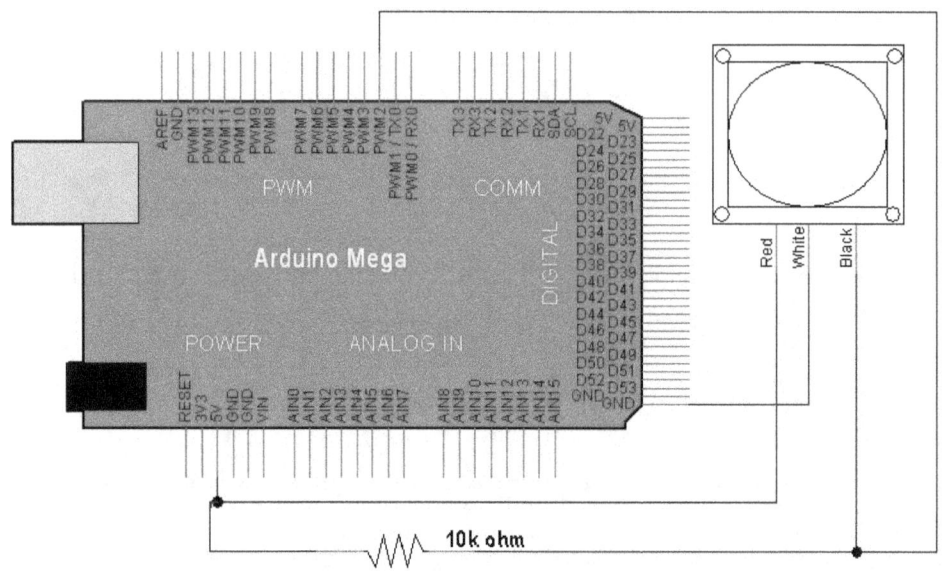

Figure 8. 6: Functional diagram for connecting the Infrared sensor to the Arduino. Note that the black wire is not ground!

➢ Build the circuit shown in Figure 8. 6. The right most pin in the figure is connected to an OpAmp that swings from 0 to VDC when the sensor detects a change in the infrared flux shining on the sensor.

- ➢ Connect the output to the serial monitor and open the sketch **File: Sketchbook: IRsensor**.
- ➢ Output your data to PLX-DAQ and plot it. Note the sketch is designed to output a binary signal. Either a signal change is detected or it is not.
- ➢ Once you have it working try moving your hand across the front of the detector while monitoring the output.
- ➢ Investigate how much motion, and how far from the sensor will trigger the sensor. It should see a person's body move at a couple of meters.

Part IV: Three-axis Tilt Sensor (Accelerometer)

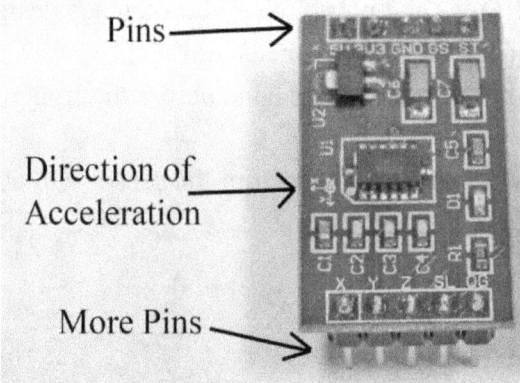

Figure 8. 7: Three-Axis Tilt sensor. Note that this sensor can detect both movement and angle of tilt because it can sense the acceleration of gravity.

The three-axis tilt sensor consists of three miniature quartz beams constructed as a capacitor and monitored with an electrical circuit. The three sensors are oriented to be mutually perpendicular to each other. If the device is moved (accelerated) along one of these axes the beam is deflected like a diving board and the capacitance changes. The resulting signal is amplified via OpAmps to produce a signal that is proportional to the acceleration.

MMA7361	Pin #
Sleep (SL)	13
Self Test (ST)	12
Zero G (0G)	11
G Select (GS)	10
X	A0
Y	A1
Z	A2
3.3	3.3V and AREF
GND	GND

Table 8. 1: List of pins to be connected from the Accelerometer to the Arduino.

- ➢ Connect the sensor pins to the Arduino as listed in Table 8. 1. Notice that the accelerometer will, when connected to the breadboard, cover all but one row on one side. You can either place jumpers beneath the accelerometer, or hang it off the side with female jumpers on the remaining pins.

- Open the sketch **File:Examples:AcceleroMMA7361: G_Force** and run it. You should see an output of X, Y, and Z values in hundredths of G in your serial monitor.
- Alter the sketch to output to PLX-DAQ and plot the acceleration on each axis as a new trace. Tilt the sensor on each axis to see each go to a min or max value.
- Test the limits of the sensor by accelerating it any way you can find that doesn't destroy it.

Part V: Ultra-sonic Range Finder

Figure 8. 8: The Ultrasonic Rangefinder consists of a speaker to send sound and a microphone to receive the echo. Our version has four pins, but more expensive versions have only one speaker/microphone and three pins.

The Ultrasonic Rangefinder operates through sonar to bounce sound waves off objects and, from the time it takes to hear a return, calculates the distance to that object. These come in several different models of increasing sensitivity and price. The model we have is the cheapest and least sensitive. The sensor can still sense a large, relatively flat object at a distance of ~15 feet if it is perpendicular to the sensor. These sensors are easily fooled, however, if the surface is at an angle to the sensor and reflects the sound waves to the side of the microphone.

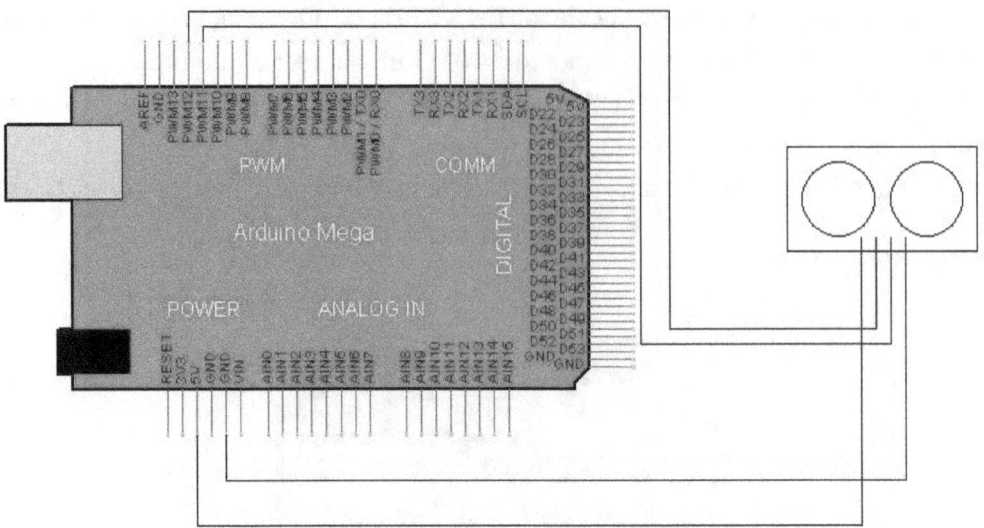

Figure 8. 9: Functional diagram for connecting the Ultrasonic Range Finder to the Arduino.

- ➢ The four pins are (from left to right) VCC, Trigger, Echo, and Ground. Connect the VCC pin to the 5VDC and the ground to the GND pins on the Arduino.
- ➢ Open the sketch **File: Sketchbook: Ping**, and see which pins are defined for the Trigger and Echo pins on the sensor. Connect the Trigger and Echo pins to those pins listed as shown in Figure 8. 9.
- ➢ When the Arduino strobes the Trigger pin the device produces a short ultrasonic pulse via a piezoelectric speaker. The Arduino then samples the Echo pin for a detected return signal on the second piezoelectric microphone. When the signal is detected, the time difference is then measured and distance calculated given the speed of sound.
- ➢ Connect the device using the jumper wires and download the sketch. Monitor the output with PLX-DAQ. Try using the sensor to measure the distance to something like a cardboard sheet.
- ➢ Once you have it working, try taking an average of values to improve precision and alter the sketch, if necessary, to improve the calibration of the sensor. The true distance to an object can be found with a tape measure.

Part VI: Multiple Sensors (Ultrasonic and Accelerometer)

The Arduino Mega is perfectly suited to run multiple sensors at the same time due to its faster processor, larger cache, and expanded number of I/O pins over earlier models. You will now set up two sensors to measure at the same time to measure the periodic motion of a weight on a spring.

- In the lab, there is equipment you will need for this setup.
    - Heavy Bi-Directional Clamp
    - Long Metal Pole
    - Heavy Spring
    - Disk Shaped Weight
    - Light Bi-Directional Clamp
    - Claw Arm
- Set up the equipment so that the weight is attached to the bottom of the spring and allowed to oscillate up and down from the claw arm attached to the top of the pole.
- You will place the Ultrasonic Rangefinder beneath the weight to measure the distance of the weight from the floor. The small vise grip at your station can be used to hold the rangefinder vertical.
- The Accelerometer will be placed on the weight to directly measure the acceleration of the weight through the oscillation.
- To have two sensors in two places with one Arduino, you will need to use a length of cable to run to either the rangefinder or the accelerometer on a small breadboard. These lengths of cable can be found in cabinet "C", and the breadboards can be found beneath drawer "A". I will leave it up to you to decide which sensor you want to have at the end of the cable.
- Have both sensors read out to PLX-DAQ.
- Calculate the derivative and double derivative of the position vs. time from the rangefinder in Excel.
- Graph these quantities on the same graph with the acceleration to show that the double derivative is the acceleration of the motion. You may need to do a vector sum of the values from the accelerometer if it is tilted such that the total acceleration is split among two or three axes.

Summary

In this lab, we have examined a range of sensors that can be sampled using the Arduino. These sensors should give some ideas of what is possible with the Arduino and help you to develop ideas for your class project.

Lab 9 - Reference Voltages and Multiplexers

Goals: Investigate the effects of different reference voltages available on the Arduino Mega, and how to improve the accuracy and precision of sensor readings.

1. Internal Reference Voltages
2. External Reference Voltages
3. Precision Voltage Supplies
4. Multiplexer

List of Equipment and Parts

1. One Arduino Mega and Cables
2. Computer
3. Photoresistor x2
4. Temperature Sensor
5. Two LEDs
6. Jumper wires
7. Breadboard
8. LM4040D41ILPR Zener
9. 4051 Multiplexer
10. DC Motor
11. TIP120 Darlington Pair

Introduction and Overview

The Arduino Mega is a fantastic tool for interfacing between sensors and computers, but it is limited to its 10-bit analog to digital converter (ADC). The 1024 steps divided from 0 to 5V results in a precision of 4.88mV/step. While this may be adequate for this class and for most casual applications, for professional work it will generally be too coarse.

We can take several steps to help mitigate this limitation of the Arduino. The Arduino Mega is special compared to the other models in that it offers four different means of providing a reference voltage for its ADC. In addition to the default 5V reference, it has two internal references, the 1.1V and 2.56V, as well as an external reference voltage pin.

Multiplexer

The Multiplexer is an IC that can be many sizes including 4 to 1, 8 to 1, and 16 to 1 as common types. In each case, the IC has 4, 8, or 16 pins that can be connected to either inputs (i.e. sensors) or outputs (i.e. motors). Inside is a switch that connects one of these pins to just one pin that can be connected to either an Arduino analog pin to read in values, or a DC voltage to power outputs. Which pin is connected to the Arduino is addressed through 2 to 4 special pins that read in binary values (HIGH or LOW). In the case of the 8 to 1, there are three pins where LOW-LOW-LOW is the first, and HIGH-HIGH-HIGH is the last pin.

Multiplexers can then allow one analog pin to read in 4, 8, or 16 sensors addressed by 2, 3, or 4 digital pins. Each sensor can only be read after the time interval it takes the Arduino to switch digital pins from high to low. If there are 16 sensors on one pin, then you can only sample them $1/16^{th}$ as fast as a single sensor. Also, multiplexers can generally only handle small currents of less than 100mA.

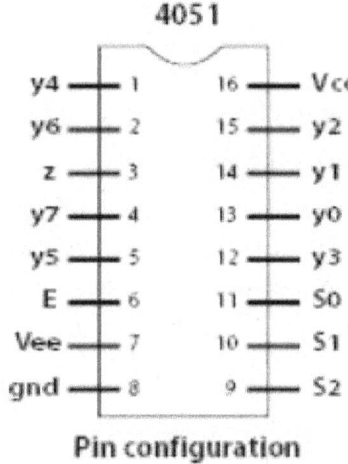

Figure 9. 1: The 4051 Multiplexer/Demultiplexer. It has 8 I/O pins and is fast switching.

The Multiplexer we will be using is the 4051 8 to 1 multiplexer/demultiplexer that can be seen in Figure 9. 1. The pin configuration is as follows:

Vcc	Positive Power Supply (5V for our use)
Vee	Negative Power Supply (Ground for our use)
gnd	Ground
E	OFF/ON for the IC. Connect to Ground for ON; Connect to 5V for OFF
Z	I/O pin connected to either Power or to the Arduino
S0,S1,S2	Addressing pins: HIGH or LOW
y0 through y7	I/O pins connecting to sensor or outputs

Procedure

Part I: Operating Reference Voltage

Up until now, whenever you have measured a voltage with the Arduino, you have used the board's operating voltage of ~5V as a reference voltage. The Arduino divides the gap between zero and 5V into 1024 equal steps of voltage. Any incoming voltage is then assigned the step number corresponding to which of the 1024 steps is closest to the incoming voltage. For low precision and low accuracy measurements, this approach works, but if you want to use the Arduino for semi precise measurements, this is the least advisable method.

The operating voltage of the board suffers from the problem that it is not well regulated. It can fluctuate by at least 10% during normal operation. Additionally, the board voltage is heavily dependent on the source of input power for the Arduino. A board run from a wall socket will have a higher operating voltage than one run from the USB of a computer.

You will now measure this difference.

- Build the correct circuit to connect the light sensor to the Arduino, but power the circuit with your 5V DC power supply instead of the Arduino.
- Plug the Arduino into the computer via the USB and sample the voltage across the light sensor in a steady ambient light. Record an average value seen by the Arduino.
- Then, while you are sampling with the Arduino, plug it into the wall with the AC adapter.
- Record the new average value seen by the Arduino and compare to the old value.
- Finally, disconnect the Arduino from sampling the light sensor and measure the output voltage with the DMM.
- From your measurements, determine the reference voltage used by the Arduino when run by the USB vs the AC adapter and find the percentage drop/increase between the two.

Part II: Internal Reference Voltages

There are two other internal reference voltages that are available for use with the Arduino, 2.56 and 1.1 Volt. These alternative references hold at least one advantage over the default 5V, they offer higher resolution to your measurements as long as the input voltage is lower than the reference voltage.

The default 5V, split into 1024 equal steps offers a resolution of 4.88 mV, where the 2.56 and 1.1 offer resolutions of 2.5 mV and 1.07 mV respectively. So, as long as your input voltages are small enough, you gain double or even quadruple the resolution by setting the reference voltage lower.

- Use the same setup you had previously for the light sensor, but change the voltage divider such that the voltage output for ambient light is ~2.0 V. Also, power the Arduino from the USB only.
- Add in the Setup of your sketch the line that reads: **analogReference(INTERNAL2V56);**
- Record the size of the step in the data you see from random fluctuations. Does this match the calculation in the step above?
- Plug in the AC adapter to the Arduino and re-examine the data. Did it shift the same as in Part I? Does it change the size of the data resolution?
- Repeat the above steps, but change the reference to: **analogReference(INTERNAL1V1);**

Part III: External Reference Voltages

As you have seen in the previous parts, even though you can add higher resolution to you data collection, the values can shift based on the variations in the internal reference voltages, no matter which value you choose. A solution to this issue is to use an external power source, one that is invariant with time and temperature.

There are many high precision reference voltages that are available for sale. One such device, the LM4040D41ILPR (from TI) zener diode is cheap and easy to use. Simply connect the diode as shown in the diagram and any voltage in parallel with the zener will maintain the same voltage as the breakdown voltage (4.096 in this case). This is no different from other zener diodes, but this one has an accuracy of 0.1%. Also, you can choose this diode to be in a variety of voltage outputs to suit your need.

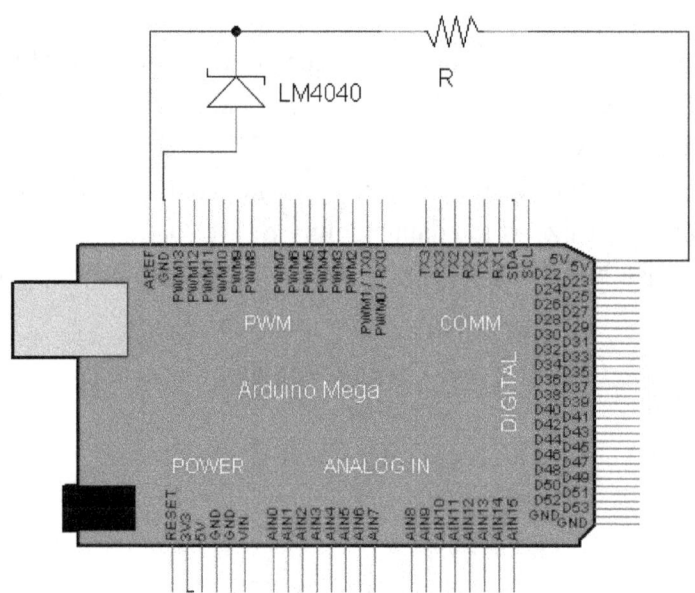

Figure 9. 2: The 5V supply on the Arduino feeds a voltage divider where the second resistor is a zener diode. The AREF pin is in parallel with diode which maintains a constant voltage. Any voltage fed into the AREF pin must be below 5V and 40mA!

- ➢ To determine the value of the resistor, you must calculate the range of total current that will be required. The AREF pin on the Arduino has an internal resistance of 32kΩ. Using Ohm's law, and the breakdown voltage of the zener diode, calculate the current that will flow into AREF.

- ➢ The diode has a minimum and maximum current that must be considered. The LM4040D41ILPR has a minimum of 73μA and a maximum of 15mA. Add the minimum current of the diode to the current found above to get the total current through R.

- The minimum voltage through the USB port to the Arduino is 4.4V. Since we know that the diode will have a 4.096 voltage drop, the voltage drop across the resistor, R, can be calculated as the difference between 4.4 and 4.096, or 0.304V. Using this voltage, the total current found above, and Ohm's law, calculate the resistance of R.

$$$$

This value of R is the value at which, even under the worst conditions, the zener diode will have enough current to activate.

If you do not have an LM4040, another choice is to take an existing power supply and add a voltage regulator to smooth out any variations in the output. We have the LM317 adjustable voltage regulator, and it can be used as described at the end of Lab 3.

If you do not have access to a voltage regulator, the 3.3 volt line on an ATX power supply is well regulated and will supply a much more precise and accurate reference voltage than the Arduino can supply.

- If you have the LM4040 available, build the circuit shown in Figure 9. 2.
- If you do not have the LM4040, the 3.3 line from the power supply will suffice. Choose a resistor to place in series with the power supply such that the current is well below 40mA. BE ABSOLUTELY CERTAIN that the voltage being fed into the AREF pin is below 5V and is below 40mA of current. Feeding too much power into the AREF pin is one of the sure ways of destroying the otherwise rugged Arduino.
- If you use the power supply, connect the ground of it to the ground pin the Arduino so that they share a common ground.
- Modify the previous sketch you were using to use the external reference voltage: **analogReference(EXTERNAL);**
- Test the new reference voltage by measuring the ambient light with just the USB to the Arduino as well as with the AC adapter plugged into the Arduino. Does this method fix issues encountered above?

Part IV: Multiplexer Input

Another way to expand the number of I/O pins would be through the use of a multiplexer. The one we will be using is the 4051, 8 to 1 multiplexer/demultiplexer shown in Figure 9. 3.

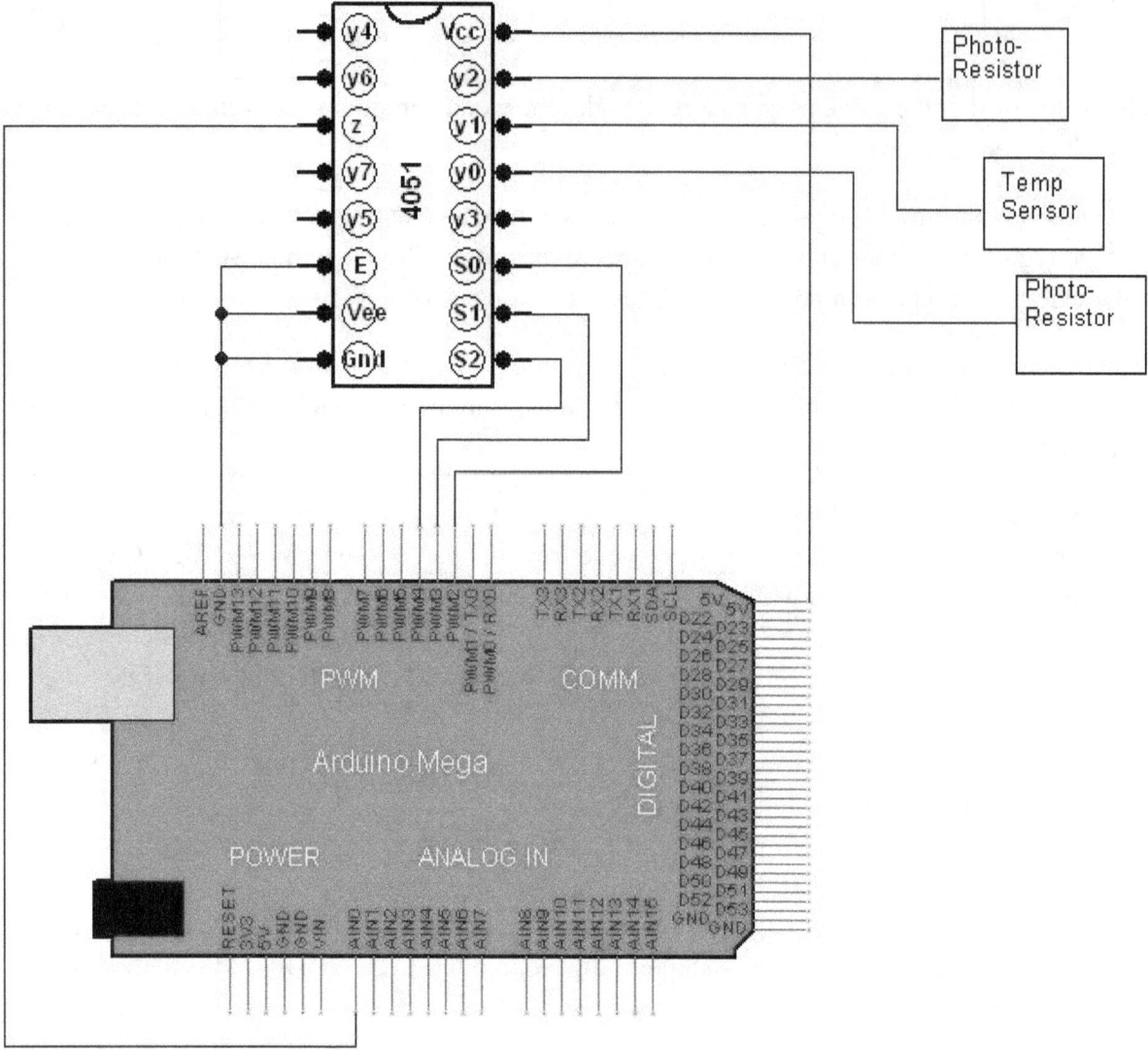

Figure 9. 3: The 4051 multiplexer connected to the Arduino controlling three different sensors. Pins 2, 3, and 4 address which sensor is connected to the analog pin on the Arduino.

- Set up the circuit shown in Figure 9. 3. If you have trouble connecting the three sensors appropriately, review Lab 8.
- Load the sketch **File:Sketchbook: ReadMultiplexer2Excel**. Notice how the Arduino addresses the multiplexer with pins 2, 3, and 4. It uses binary code for 0 through 7 where the individual bits control whether each pin is high or low.
- This sketch will only analog read the values from each sensor. You will need to use the calibration from Lab 8 to output a real data value for the temperature sensor. Once the sketch is working, save it and move on to the next step.

Part V: Multiplexer Output

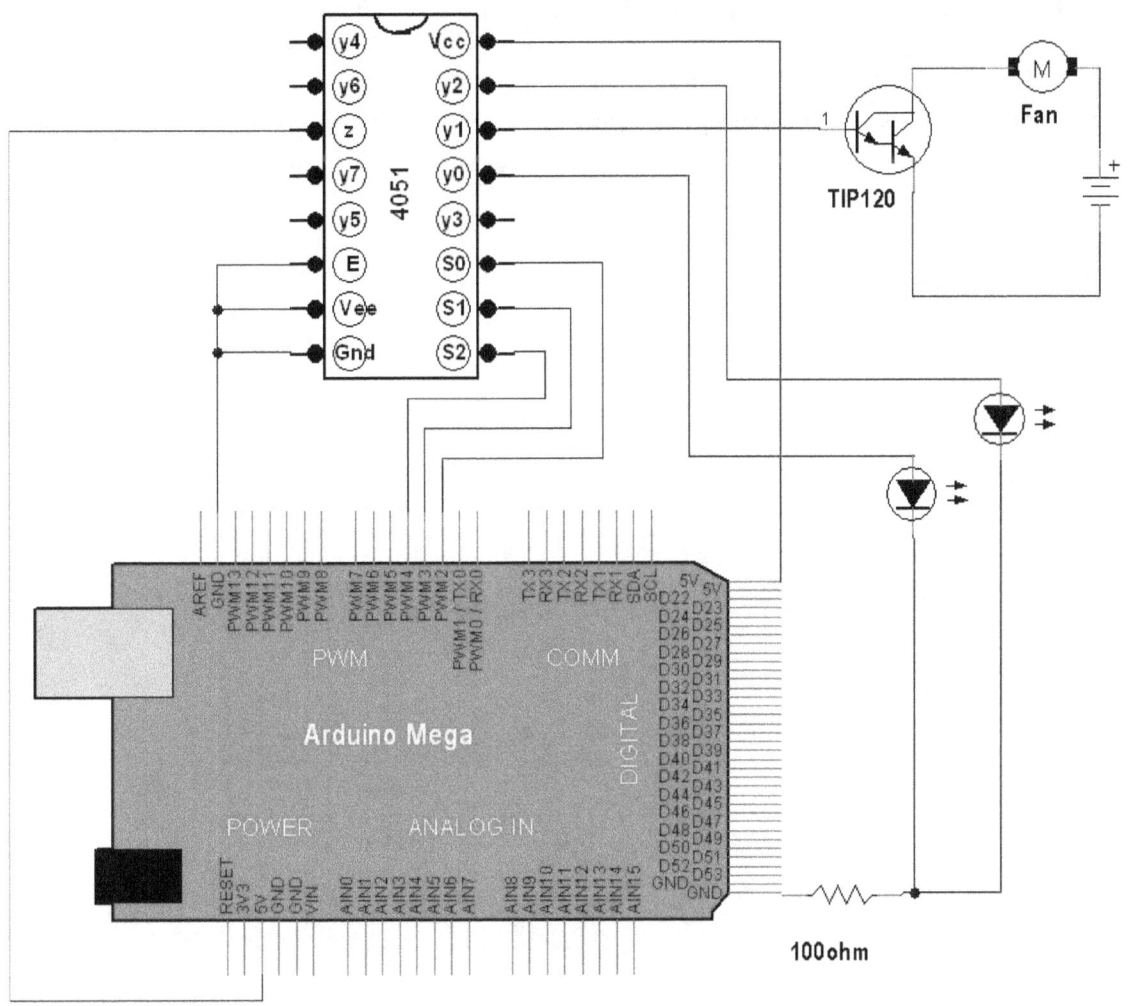

Figure 9. 4: The Arduino controlling three outputs (two LEDs and a fan) through the 4051 multiplexer. Notice the analog pin connection to z has been replaced with a 5V output to power the LEDs and fan.

- Setup the circuit shown in Figure 9. 4. Choose a case fan that requires more than $100mA$ of current and 12V. Use the TIP120 Darlington pair transistor to control the fan. Your DC power supply will power the fan with 12V through the transistor. The TIP120 can handle large currents to the base, and we are simply using it as a switch so there is no need to build the AC amplifier circuit around the transistor.
- Open the sketch **File:Sketchbook: OutputMultiplexer** and run the sketch. It should cycle through the LEDs and the motor, turning on just one output at a time.
- Modify the sketch so the multiplexer only switches to the three pins in use and remove all delays. Though the Arduino can switch pins in about $10\mu s$, the total time each component has power is only about 33% of the total. Do the LEDs appear dimmer? Does the fan run slower?
- Add one Serial.println(); command to the sketch and observe the effects again.

Lab 10 – The SPI and I2C bus

Goals: Investigate the properties of the SPI and I2C bus.

1. Connect two Arduinos through the SPI and I2C buses
2. Understand how to pass commands from one Arduino to another
3. Understand how to pass sensor reading from one Arduino to another
4.

List of Equipment and Parts

1. Two Arduino and Cables
2. Computer
3. Two Momentary Pushbuttons
4. Temperature Sensor
5. Two LEDs
6. Jumper wires
7. Breadboard
8.

Introduction and Overview

While the Arduino Mega has a larger number of I/O pins compared to other models, it is still limited to 50 pins, of which, only 15 are analog pins. Monitoring nearly more than 15 different sensors is possible but would require a large number of wires all running to the Arduino. Additionally, transmitting and reading small voltages from many sensors on wires of various lengths presents the challenge of signal degradation over long cables. The solution to both problems is to use a communication bus that converts the analog signal to a digital one that can be transmitted over few wires without alteration of the data in the signal. The two communication buses we will focus on are the SPI and I2C bus.

SPI

The SPI bus (Serial Peripheral Interface) or SSI (Synchronous Serial Interface) allows for the control of and communication with multiple devices through three wires for all plus one extra wire for each device. SPI is a synchronous serial data link that is an industry standard for high speed communication over short distances. It's extremely fast communication speed is ideal for reading large amounts of information over short distances. Indeed you have already used it when connecting the SD card shield to the Arduino.

The SPI bus uses a three wire system plus one for communication as shown in Figure 10. 1. There is only ever a single master but there can be multiple slaves connected to the master. All slaves share the SCLK, MOSI, and MISO lines, while each slave receives its own SS lines.

- SCLK: Serial Clock on Mega Pin 52; output from master and is used to keep all communication synchronized. Can also be seen written as SCK or CLK.
- MOSI: Master Output, Slave Input on Mega Pin 51; output from master and is the signal sent to the slaves. Can also be seen written as SIMO, SDO, DO, DOUT, SI, or MTSR.
- MISO: Master Input, Slave Output on Mega Pin 50; output from slave and is the response sent back to the master. Can also be seen written as SOMI, SDI, DI, DIN, SO, or MRST.
- SS: Slave Select on Mega Pin 53; the wire is held LOW to activate the slave device for communication and held HIGH to deactivate the slave. Can also be seen written as nCS, CS, CSB, CSN, nSS, STE, SYNC.

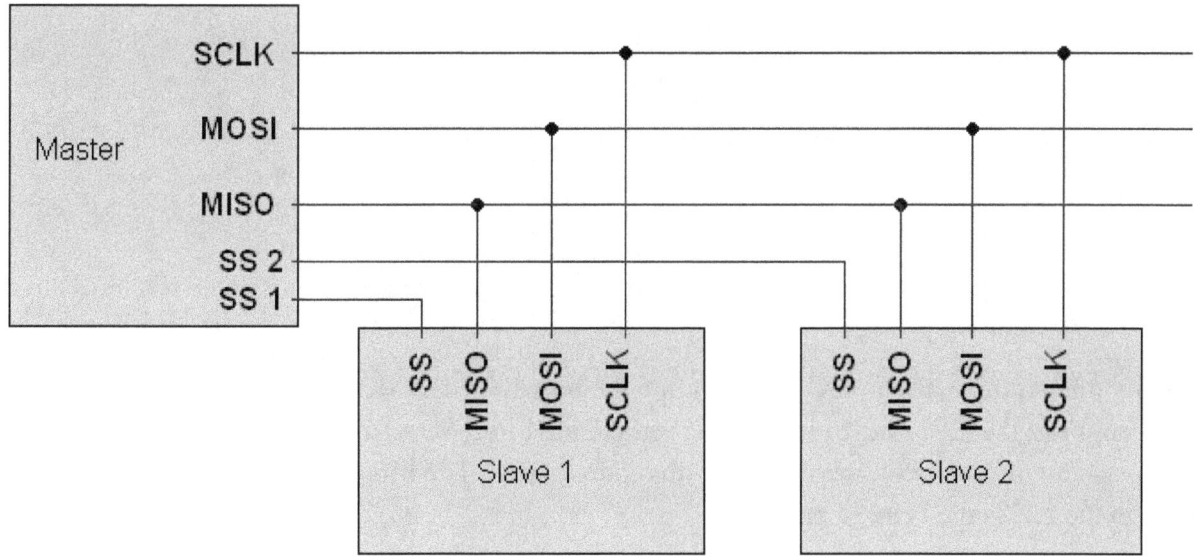

Figure 10. 1: A diagram showing the connections of one master and two slaves through an SPI bus. Notice that each slave requires it's own SS connection.

Because of the clock signal, the SPI bus has the advantage of being very fast communication (MHz range), ideal for large amounts of information. It can also transfer the same command to many slaves in unison. It is also an industry standard that the Arduino uses for many of its shields, including the SD card, Wi-Fi, and Bluetooth shields. The disadvantages are its short range, becoming ineffective at distances larger than one foot. Also, unlike many other communication protocols like the I2C, it requires three wires plus one extra per slave. This dramatically restricts its usefulness for communication with large numbers of sensors.

I2C

The I2C bus (Inter-Integrated Circuit) allows for the control of multiple devices through just two wires. Each device is given a 7-bit address to distinguish amongst the devices. What this means for the Arduino, is that up to 127 sensors could be connected and control through just two pins with the limitation that the total length of the wire connecting all the devices must be less than about one meter. Also, the I2C bus is much slower than the SPI bus, operating in the kHz range.

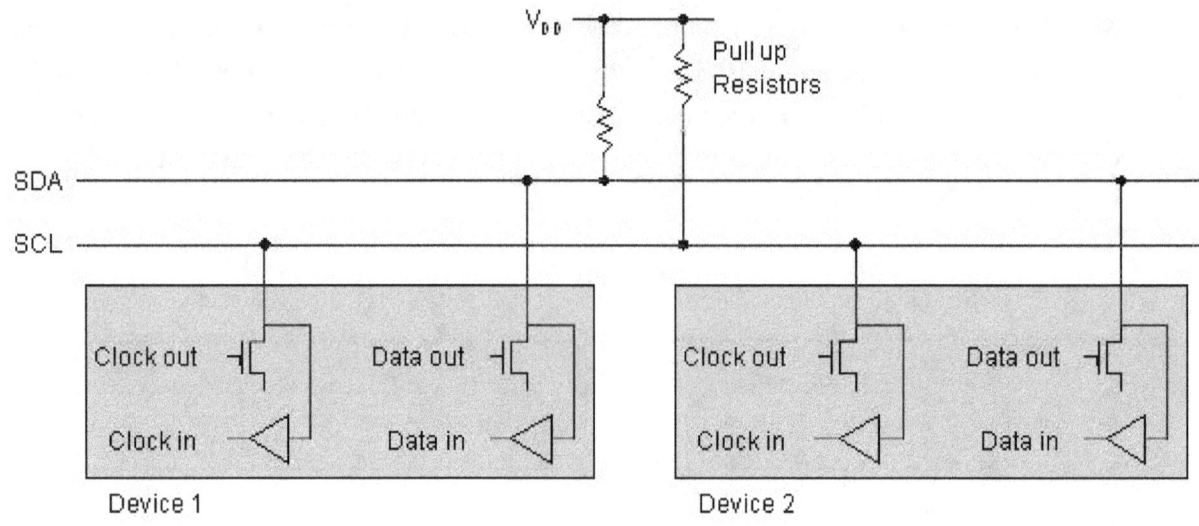

Figure 10. 2: Block diagram of an I2C bus connecting two devices. Notice the pull up resistors that are necessary for data transmission.

Luckily, the Arduino has an I2C bus built into the board. On the Mega, the pins 20 and 21 are the connections to the I2C. These pins are the Serial Data Line (SDA) and the Serial Clock Line (SCL) and are used for communication with other devices. The library used to communicate through the I2C is the Wire library.

Because our sensors do not have the I2C protocol built in, we will need to use a slave device to interface between the I2C and the sensors. The slave device will be a second Arduino. If your group does not have access to a second Arduino, you may need to share with another group.

Procedure

Part I: SPI Bus – One Master and One Slave

We will use two Arduinos, one master and one slave. The master will send a command to the slave telling it to execute some action. In this first example, the command will be a pushed button, and the Master will send a text to the slave which will then light an LED.

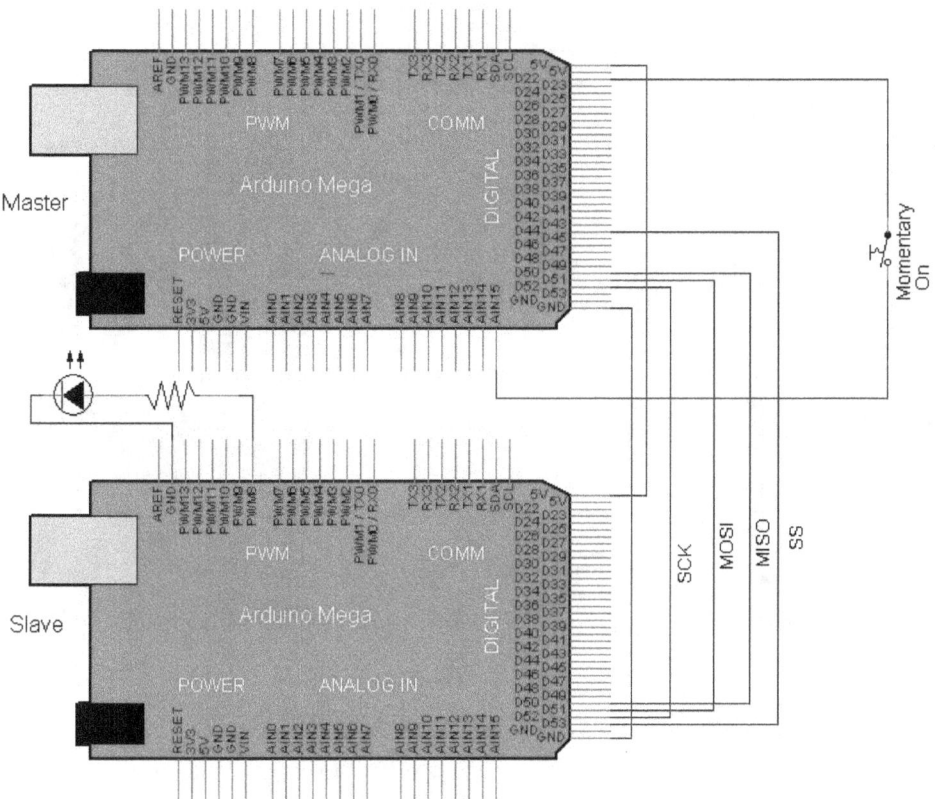

Figure 10. 3: The wiring diagram for a simple Master controlling one Slave connected through the SPI bus. The Master is the upper Arduino that, when the button is pressed, the Slave lights up the LED.

- Set up the two Arduinos as shown in Figure 10. 3. Notice that, in the SPI configuration, MOSI connects to MOSI and MISO connects to MISO.
- If the 5V and Grounds are connected, you will need to only plug one Arduino into the USB for power and it will power both boards.
- Load the sketch **File:Sketchbook: SPI_Master_Button** onto the master Arduino.
- Load the sketch **File:Sketchbook: SPI_Slave_LED** onto the slave Arduino. Note that the slave needs no specific address for the master to communicate with it, because the SS pin will be set to "LOW" when the master wants to communicate.
- Upload both sketches to their respective Arduinos. Leave the Slave connected to the USB and start the Serial Monitor. Push the button on the Master and the slave should light the LED when it receives the text.
- Note that you could connect many slaves to one master and send the same command to all slave in unison by pulling all SS pins to low simultaneously.

Part II: I2C – Bi-directional Communication

While the SPI bus does have the capability to send information back from the slave to the master, I have found that bidirectional communication is easier to accomplish with the I2C bus. We will use two Arduinos where both boards can send and receive data. We need only distinguish between them by giving each a unique address (0-127). Also, notice that there are only two wires necessary for I2C communication.

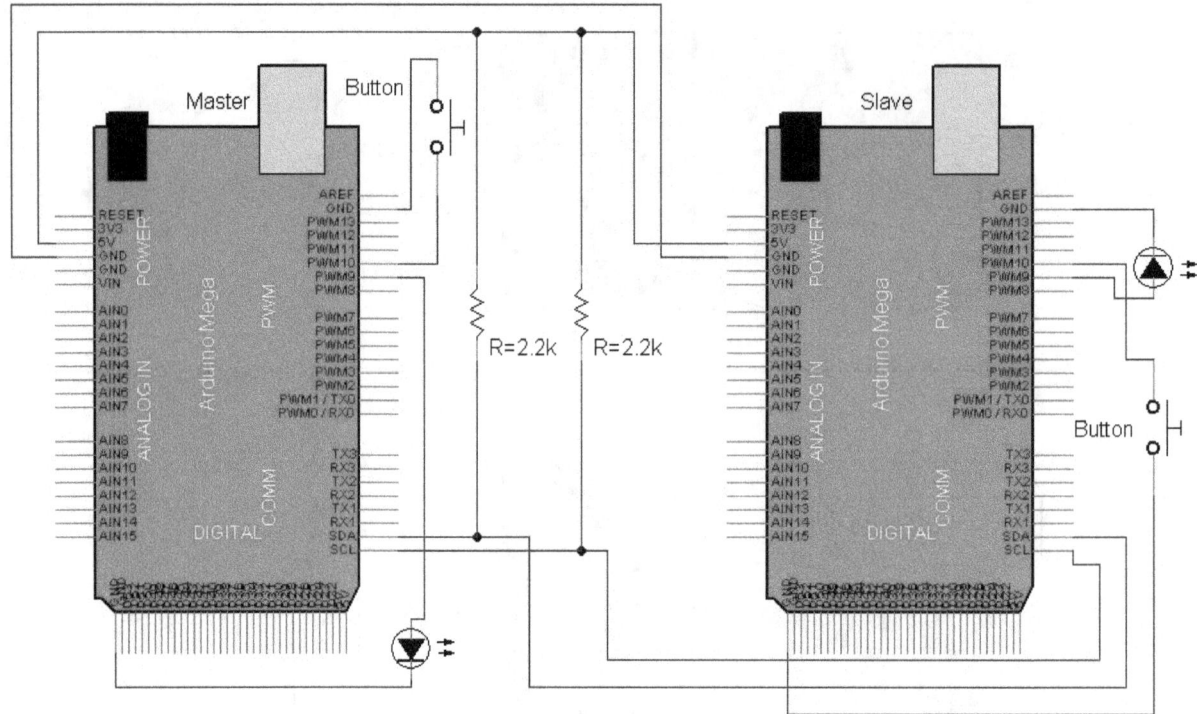

Figure 10. 4: Setup for both Arduinos to be both a master and a slave. Each button activates the LED on the other Arduino.

> Setup the Arduinos as shown in Figure 10. 4. This is similar to the previous setup, except that each Arduino has both a button and an LED. Also, I2C communication requires pull-up resistors. We will use 2.2kΩ, but consult the spec sheet for other devices as I have seen values anywhere from 1.5kΩ to 40kΩ.
> Load the sketch **File:Sketchbook: I2CMandS_Button2LED** onto both Arduinos. For the second Arduino, switch the THIS_ADDRESS and OTHER_ADDRESS values (8 & 9). Because the Arduinos will be communicating back and forth, each Arduino requires a unique address.
> Run both sketches. When you push the button one Arduino, the other LED turns off.
> Once you've gotten this sketch to work, alter it so that one Arduino sends commands to the other automatically every 100ms and not when you push a button. The result should be a steady blinking LED on the second Arduino. The button on the second Arduino should still turn off the LED on the first Arduino.
> Lower the delay between each command 10ms at a time until the LED stops blinking. Notice that this is not because the LED is blinking too fast, but because the program has

frozen. Test that this is the case by pushing the button on the second Arduino. The first Arduino LED will not turn off.
- Finally, remove all the Serial.print commands from the second Arduino sketch. The second Arduino should have no trouble receiving the commands and blinking the LED now.
- This shows that if the Arduino is using the Serial communication while receiving a command, the program can freeze.
- Finally, restore the Serial.print commands so the Arduinos communicate properly again. Place a very long delay, at least a few seconds, in the ReceiveEvent function in such a way that the LED should not light immediately after you press the button.
- Test your changes. Does the LED wait to light?
- The LED should still light immediately. The ReceiveEvent acts as an interrupt and will not execute delay commands.

Part III: I2C – Query and Receive Data

If you want to use the I2C bus as a means of data collection, one Arduino will need to query the second Arduino, which will then send data collected from a sensor.

- Keeping the same setup as in the previous section, connect a temperature sensor to A0 of the slave Arduino.
- Load the sketch **File:Sketchbook: I2CMaster_Query_Receive1** on the master Arduino.
- Load the sketch **File:Sketchbook: I2CSlave_TempRead1** on the slave.
- Notice that the Slave sketch now contains an Event function, "onRequest", that is defined in the Setup. The Master no longer just sends a number to the Slave, it sends a request for the specific number of bytes it wants which triggers the Slave onRequest event. There are two major reasons why you want to use this method when collecting data.
    - Requests from the Master are received and stored in the buffer of the Slave which is then read at the beginning of the next run through of the main loop. The bytes sent to the Master are then the specific amount requested. The whole process is efficient and results in near zero loss of data.
    - As you saw in the previous example, if the Slave is busy when the Master sent data, then the sketch may freeze. With this method, if the Arduino is busy, the command stays in the buffer until the Arduino restarts the loop. The result is that the sketch never freezes.
- Upload and run both sketches. When you press the button, the Master sends a request for eight temperature measurements, and the Slave will read from the sensor and light the LED.
- Open the serial monitor when connected to the COM port of the master. When you press the button, the requested data will be sent back to the Master, and you should see the temperature reading (uncalibrated) from the slave temperature sensor printed out.

There are many limitations to the I2C communication. The system communicates in bytes and is limited to 16 or less for any one request. Since a byte is this case is limited to an 8-bit number, the range of numbers that can be sent in one byte is 2^8 = 0 to 255. These all must be integers and cannot include decimals.

As you can see, the Master is printing decimal numbers. The sketches accomplish this by splitting the data at the Slave side into the integer value and the decimal value of each measurement. The decimal value is then multiplied by 100 so that it is one byte of data. The two subsequent bytes are then sent to the Master which then reverses the process and combines the data. This is why only eight measurements are taken, while 16 bytes of data are transmitted.

Also, note that the sketch purposely uses float data types for the temperature measurements which automatically truncates the decimals to the hundreds place. This is key, because the byte containing the decimals must be less than 255.

- Modify the code to take 16 temperature measurements and transmit them to the Master. Make two requests of the Slave for data. The first 16 bytes should contain the integers of the temperature while the second 16 bytes should contain the decimals of the temperature. You will need to also modify the Slave sketch to accept the two 16 byte transmissions, store them separately, and recombine the data.
- There is another way to send data, by sending individual bits in binary and converting the code to a number. We won't get into this here, but we will see an example of this in the next section.

Part IV: I2C: Connecting to a Sensor with inbuilt I2C

You need not connect only to the Arduino with I2C communication. I2C is very common on sensors, especially sensor modules that are meant for the Arduino. Below is an example of such a sensor, bought from Adafruit, the MCP9808 temperature sensor. Notice the SCL and SDA pins, this marks the connections for the I2C bus.

Figure 10. 5: The MCP9808 temperature module sold by Adafruit. It has an in built I2C bus and 13 bit ADC. Pull-up resistors are provided on the board. Up to 8 sensors can be addressed on a single I2C bus.

Besides the I2C bus, it also has a built in Analog to Digital Converter (ADC). This means that the resolution of the temperature measurements can be as high as 0.0625 C. It also has three address pins that, when connected to either ground or power, provide a 3 bit address for up to eight sensors on a single I2C bus. Also, an Alert pin is provided so that, when properly programmed, the sensor will send data only when the temperature passes a value of your choosing.

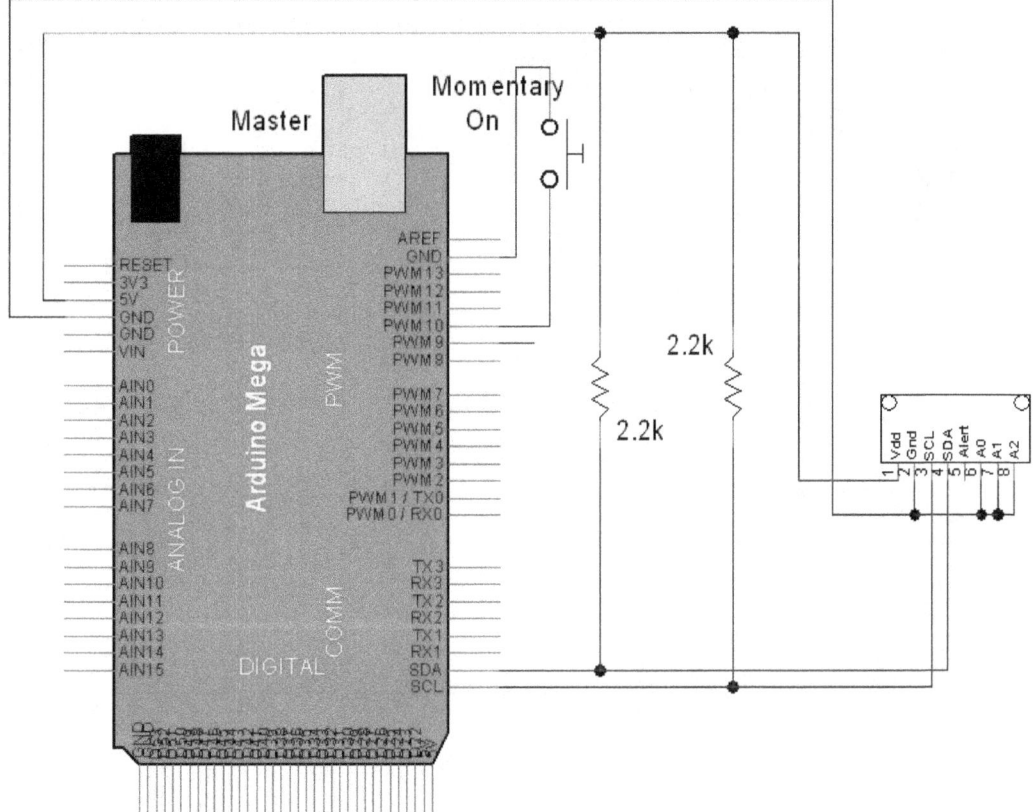

Figure 10. 6: Using the I2C bus to communicate with the temperature module, MCP9808.

- Connect the MCP9808 sensor module to the Arduino as shown in Figure 10. 6. Be sure to tie A0, A1, A2 to ground so the sensor will use the default address provided in the sketch.
- Load up the sketch **File:Examples: Adafruit_MCP9808: mcp9808test**. This sketch and the library that it uses were supplied by Adafruit. This means it already has the calibration pre-programmed and outputs real temperatures to an accuracy of 1% at room temperatures.
- Once you are satisfied that you are getting real temperatures from the module, connect the A0 pin on the module to Vdd to change the address. See if the Arduino can still communicate with the sensor.
  - Look for the line in the code that reads, "if (!tempsensor.begin()) {"and add the hex number 0x19 in the parenthesis next to "begin". This should be the new address of the sensor. Upload the new code, and connect to the sensor module.

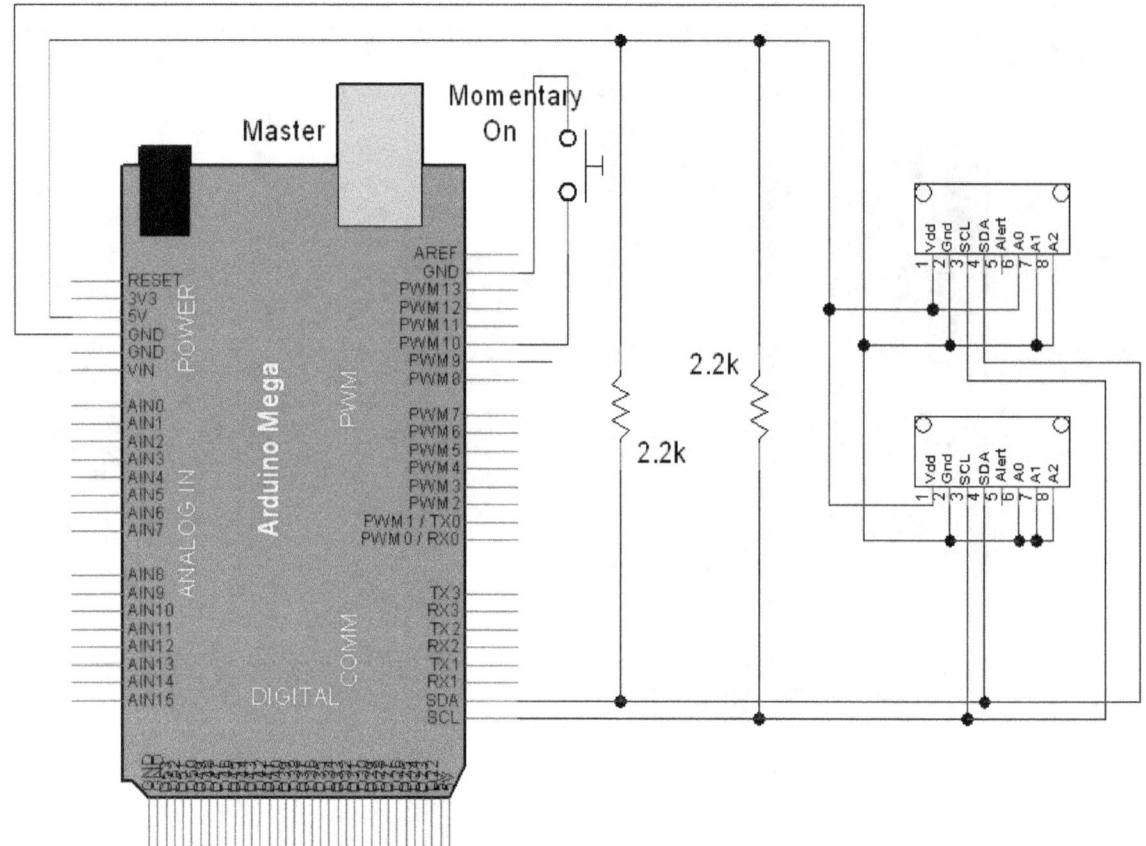

Figure 10. 7: Using the I2C bus to connect two MCP9808 temperature sensors.

- Connect another MCP9808 to the I2C bus as shown in Figure 10. 7, and connect the address pins to ground so that the new sensor module has the default address.
- Look for the line in the code that reads,
    "Adafruit_MCP9808 tempsensor = Adafruit_MCP9808();"
  Make a copy of that line, but for the copy, add the number 1 after tempsensor.
- Go the code where you changed the hex address and add another declaration of the function call with the default address, but change the function call to the same new name.
- Add the new function call in your main loop so that the Arduino queries both sensors and then prints both values.
- Upload the new code and make sure it works before moving on.

Finally, it was mentioned in the previous section that bit communication could be used over the I2C bus to communicate large numbers. Indeed you have been using this technique with the MCP9808 with realizing it.

- Go to the directory, C:\Users\Student\Desktop\arduino-1.0.1\Adafruit_MCP9808, and open the .cpp file. Look for these lines.
    - Wire.requestFrom((uint8_t)_i2caddr, (uint8_t)2);
    - val = Wire.read();

- o   val <<= 8;
- o   val |= Wire.read();
- o   return val;

You can see that the Arduino requests 8 bits of data from the MCP9808 twice. It then reads 8 bits into val. Because these 8 bits are the first of 16, the third line shifts the bits to the left to make room for the next 8. It then reads the next 8 bits and concatenates the new 8 with the first 8. All components that have built-in I2C, seem to communicate through this method, depending on how many bits are needed for the resolution of the sensor.

The conversion of this bit number to a normal byte can be seen further up in the code.

➢ Look for the lines:
- o   float Adafruit_MCP9808::readTempC(void)
- o   {
- o   uint16_t t = read16(MCP9808_REG_AMBIENT_TEMP);
- o
- o   float temp = t & 0x0FFF;
- o   temp /= 16.0;
- o   if (t & 0x1000) temp -= 256;
- o
- o   return temp;
- o   }

We can see that the variable, t, is the value read from the sensor, val, mentioned above. It is converted to a float temp with the command t & 0x0FFF. The next two lines apply the calibration to the resulting number to get real temperature. If you want to attempt bit communication between Arduinos, or communicate with another module that does not have a library, start with this library and modify as needed.

Part V: I2C Over Long Distances

I2C Bus Extender: If you need your I2C devices to be separated by a distance longer than one meter, the I2C will begin to be noisy and will fail. To fix this limitation, a pair of I2C bus extenders can extend the I2C line to greater than ten meters. More than one pair can be used on a single line to increase it to any practical length. An example is the NXP P82B715 which uses a unity gain voltage buffer boost the signal.

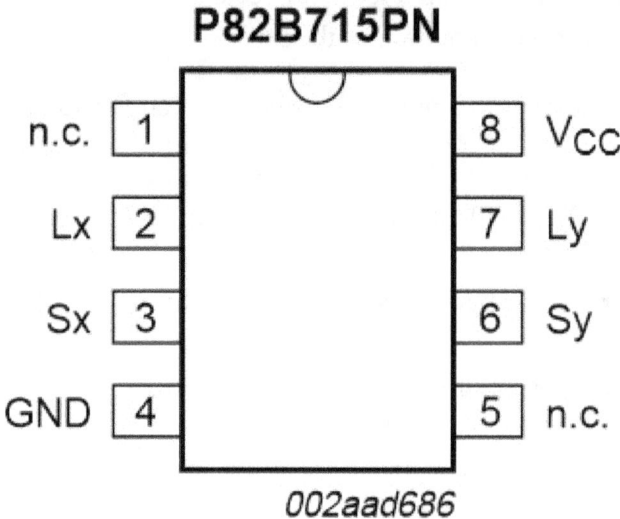

Figure 10. 8:The pinout of the I2C bus extender.

These bus extenders must be connected in pairs, one near the Master and one near the slave, to work properly. This configuration results in three segments of I2C, each of which must have pullup resistors on the SDA and SCL lines ().

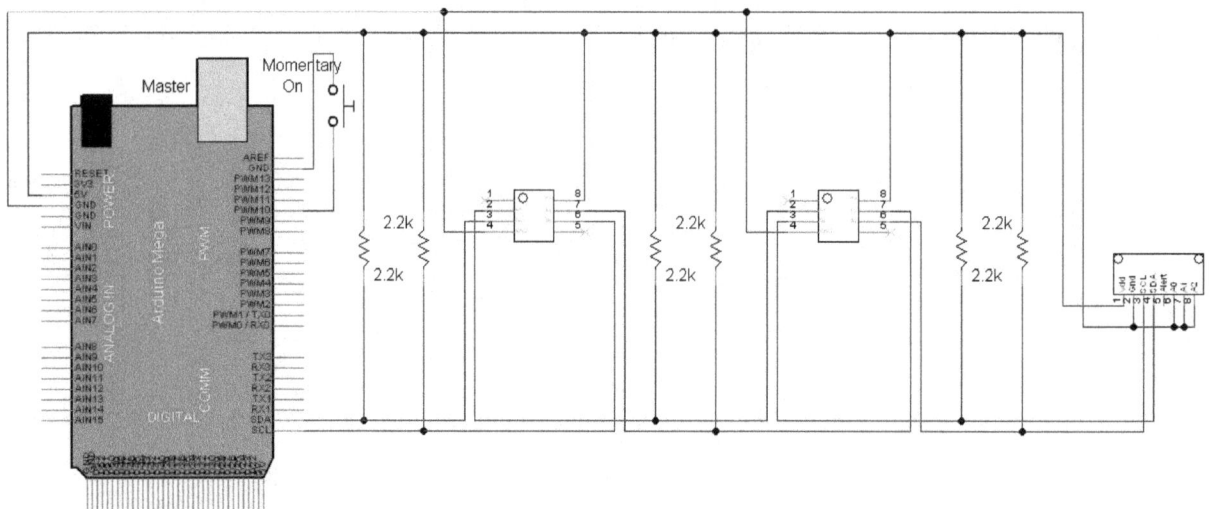

➢ Connect the MCP9808 as the Slave to the Arduino as you did previously, but use wires for the SDA and SCL lines that are longer than two meters.

- Attempt to communicate with the temperature sensor. If you still have good communication, increase the length of the wire until it no longer communicates properly.
- Connect the circuit shown in () with the MCP9808 as the Slave. Verify that it now works with the long cable.

Part VI: Software Emulated I2C

While the Arduino has built-in I2C that is easy to use, the existing library for communication makes a major assumption that the communication is perfect and no data packets get dropped. But when a data packet does get dropped, the while loops in the library will cause the program to wait forever with no timeout for the lost packet. Obviously, causing the Arduino to freeze is not an ideal situation.

One solution to this problem is a new library, I2CSoftMaster that emulates the I2C bus communication but with the added feature of a user defined timeout. The new library has the further advantage of allowing the use of any two pins to be used as SDA and SCL for the bus.

- Load the sketch **File:Sketchbook: MCP9808_SoftI2C**. Notice that the sketch no longer calls the library for the MCP9808 because the existing library is incompatible with the new I2C library. The protocols for the communication are handled completely within this sketch.
- Establish communication with the temperature sensor.

Part VII: Summary

The I2C bus is a critical means of networking Arduinos. More and more sensors have the I2C bus incorporated into it to allow for easy transmission of data and multiple sensors to connect to an Arduino with limited pins. If more I/O pins are needed, the multiplexer fills the role at the expense of sampling rate.

Part VIII: Further Applications

i. Sensor Modules with I2C: Many sensors being produced today incorporate the I2C bus for ease of connecting to microprocessors like the Arduino. These sensor modules will have an address that is unique to the model and listed in the spec sheet. If you want to connect multiple of the same sensor, this unique address is no longer unique from the Master Arduino's perspective. To help with the problem, most of these sensors leave three or less pins as inputs that act a lot like the selector pins on the multiplexer. Usually listed as A0, A1, and A2, they allow the alteration of the last bit of the address to provide for up to eight unique addresses even for the same model.

ii. I2C Level Shifter: Not all sensors with I2C buses will operate at the 5V that the Mega operates. Many are now switching to 3.3V operation and communication level. What this means is that the HIGH pulses from the I2C on the sensor will be at 3.3V which is too low for a 5V I2C bus to recognize. To solve the miscommunication, you will need to use an I2C level shifter. Placed between the two devices, the level shifter will bump the 3.3V up to 5V and the 5V down to 3.3V allowing communication normally.

iii. I2C Switch: If you find that the multiple addresses on the sensor module you have are not enough for the number of sensors you wish to connect, the I2C switch will fix that. The switch acts as a multiplexer for the I2C bus. So, if your sensor has only eight unique addresses, and your switch is 8-channel, then you can address 64 of those sensors. An example of this is the TI PCA9548A.

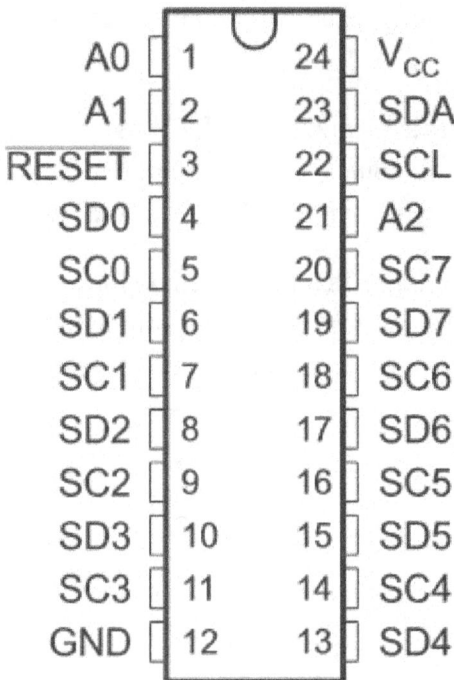

Appendix: List of Possibly Useful Websites

Java Applets for Electronics:

Ohm's Law http://www.walter-fendt.de/ph14e/ohmslaw.htm
Combining Resistors http://www.walter-fendt.de/ph14e/combres.htm
Potentiometer http://www.walter-fendt.de/ph14e/potentiometer.htm
Wheatstone Bridge http://www.walter-fendt.de/ph14e/wheatstone_e.htm
Simple AC Circuits http://www.walter-fendt.de/ph14e/accircuit.htm
RLC Circuit http://www.walter-fendt.de/ph14e/combrlc.htm
Oscillating Circuit http://www.walter-fendt.de/ph14e/osccirc.htm

Math Review:

SOS Math http://www.sosmath.com/
Review of Complex Arithmetic
    http://www.physics.uwyo.edu/~mpierce/P3640/Complex_Review.pdf

Electronics Review:

Nice Review of Electronics
    http://www.physics.uwyo.edu/~mpierce/P3640/Complex_Review.pdf

Problem Solving Strategies:

Problem Solving Strategies http://www.physics.uwyo.edu/~mpierce/P3640/Strategy.doc

Arduino Links:

Arduino Home Page http://www.arduino.cc/

Gammon Software Solutions http://www.gammon.com.au/welcome.html

Maker's Electronics Links:

Evil Mad Scientist Laboratories http://www.evilmadscientist.com/
Build Circuit http://www.buildcircuit.com/
Open Source Electronics http://www.open-elecgronics.com/
AdaFruit Industries (Great Resource) http://www.adafruit.com/
Sparkfun Electronics (Great Resource) http://www.sparkfun.com/

Additional Reference Links:

Arduino Plugin for MS Visual Studio http://www.visualmicro.com/
Arduino Index and Knowldege Base http://www.freeduino.org/
Limor Fried's (Adafruit Industries) Page http://www.ladyada.net/
The $25 Linux Computer (Wow!) http://www.raspberrypi.org/
Robotics for Kids http://mindstorms.lego.com/
Hack a Day http://hackaday.com/

Hack-n-Mod http://hacknmod.com/
A Python Tutorial http://docs.python.org/2/tutorial
Google's Python Tutorial http://developers.google.com/edu/python/
Google's Robot Operating System http://wiki.ros.org/

Electronics Tutorial Links:
Nice Electronics Tutorial http://www.electronics-tutorials.ws/

Semiconductor Physics Links:
Britney Spears' Tutorial on Semiconductor Physics http://britneyspears.ac/lasers.htm

Articles and News Links:
Wired Magazine: Just What is an Arduino?
   http://www.wired.com/gadgetlab/2008/04/just-what-is-an/
Links to the 30 Wiimote Hacks (Hack-n-Mod)
   http://hancknmod.com/hack/top-30-wiimote-hacks-of-the-web/
Hack-n-Mod's list of Top-40 Arduino Projects (cool!)
   http://hacknmod.com/hack/top-40-arduino-projects-of-the-web
Make Magazine http://makezine.com/
Make Projects Page (Very Cool!) http://makeprojects.com/

Appendix B: List of Arduino Sketches

All code listed in this section can be found on github at
https://github.com/ewood6/ArduinoClassCode

Example Code 1: 7 Segment Display – Code for controlling a seven segment display with the Arduino. It requires no additional library. Code continues on following page

```
// Arduino 7 segment display example software
// http://www.hacktronics.com/Tutorials/arduino-and-7-segment-led.html
// License: http://www.opensource.org/licenses/mit-license.php (Go crazy)
// Define the LED digit patters, from 0 - 9
// Note that these patterns are for common cathode displays
// For common anode displays, change the 1's to 0's and 0's to 1's
// 1 = LED on, 0 = LED off, in this order:

//Arduino pin: 2,3,4,5,6,7,8
byte seven_seg_digits[10][7] =
{
 { 1,1,1,1,1,1,0 },  // = 0
 { 0,1,1,0,0,0,0 },  // = 1
 { 1,1,0,1,1,0,1 },  // = 2
 { 1,1,1,1,0,0,1 },  // = 3
 { 0,1,1,0,0,1,1 },  // = 4
 { 1,0,1,1,0,1,1 },  // = 5
 { 1,0,1,1,1,1,1 },  // = 6
 { 1,1,1,0,0,0,0 },  // = 7
 { 1,1,1,1,1,1,1 },  // = 8
 { 1,1,1,0,0,1,1 }   // = 9
};
void setup() {
  pinMode(2, OUTPUT);
  pinMode(3, OUTPUT);
  pinMode(4, OUTPUT);
  pinMode(5, OUTPUT);
  pinMode(6, OUTPUT);
  pinMode(7, OUTPUT);
  pinMode(8, OUTPUT);
  pinMode(9, OUTPUT);
  writeDot(0);  // start with the "dot" off
}
void writeDot(byte dot) {
  digitalWrite(9, dot);
}
```

```
void sevenSegWrite(byte digit)
{
 byte pin = 2;
 for (byte segCount = 0; segCount < 7; ++segCount) {
  digitalWrite(pin, seven_seg_digits[digit][segCount]);
  ++pin;
 }
}

void loop()
{
 for (byte count = 10; count > 0; --count) {
  delay(1000);
  sevenSegWrite(count - 1);
 }
 delay(4000);
}
void sevenSegWrite(byte digit)
{
 byte pin = 2;
 for (byte segCount = 0; segCount < 7; ++segCount)
 {
  digitalWrite(pin, seven_seg_digits[digit][segCount]);
  ++pin;
 }
}

void loop()
{
 for (byte count = 10; count > 0; --count)
 {
 delay(1000);
 sevenSegWrite(count - 1);
 }
 delay(4000);
}
```

Example Code 2: MouseLEDs - Arduino code for controlling LEDs with the computer. Coninues on following page.

```
// Mouse LEDs for the Arduino
// This receives information from serial port (USB) from the Processing and lights or
// or unlights LEDs connected to the Arduino.

void setup() {
  // initialize the digital pins as an output.
  pinMode(2, OUTPUT);
  pinMode(3, OUTPUT);
  pinMode(4, OUTPUT);
  pinMode(5, OUTPUT);
  pinMode(6, OUTPUT);
  pinMode(7, OUTPUT);
  pinMode(8, OUTPUT);
  pinMode(9, OUTPUT);
  pinMode(10, OUTPUT);
// Turn the Serial Protocol ON
  Serial.begin(9600);
}

void loop() {
  byte byteRead;

  /* check if data has been sent from the computer: */
  while (Serial.available() > 0) {

    /* read the most recent byte */
    byteRead = Serial.read();
    //You have to subtract '0' from the read Byte to convert from text to a number.
    byteRead=byteRead-'0';

    //Turn off all LEDS
    for(int i=2; i<11; i++){
      digitalWrite(i, LOW);
    }
```

```
    if(byteRead>0){
      //Turn on the relevant LEDs
      for(int i=1; i<(byteRead+1); i++){
        digitalWrite(i+1, HIGH);
      }
    }
  }
}
```

Example Code 3: MouseLEDs – Code for Processing to capture mouse coordinates from the computer and sent it through the serial port to the Arduino. Continued on next page.

```
//MouseLEDs for Processing
//This code captures the coordinates of the mouse and sends them through the serial
//port to the Arduino.

import processing.serial.*;
import cc.arduino.*;

// Global variables
int new_sX, old_sX;
int nX, nY;
Serial myPort;

// Setup the Processing Canvas
void setup(){
  size( 800, 400 );
  strokeWeight( 10 );

  //Open the serial port for communication with the Arduino
  //Make sure the COM port is correct
  myPort = new Serial(this, "COM14", 9600);
  myPort.bufferUntil('\n');
}

// Draw the Window on the computer screen
void draw(){

  // Fill canvas grey
  background( 100 );

  // Set the stroke colour to white
  stroke(255);

  // Draw a circle at the mouse location
  ellipse( nX, nY, 10, 10 );
```

```
//Draw Line from the top of the page to the bottom of the page
//in line with the mouse.
  line(nX,0,nX,height);
}


// Get the new mouse location and send it to the arduino
void mouseMoved(){
 nX = mouseX;
 nY = mouseY;

//map the mouse x coordinates to the LEDs on the Arduino.
 new_sX=(int)map(nX,0,800,0,10);

 if(new_sX==old_sX){
   //do nothing
 } else {
   //only send values to the Arduino when the new X coordinates are different.
   old_sX = new_sX;
   myPort.write(""+new_sX);
 }
}
```

Example Code 4: Code for measuring temperature with the TMP36 sensor. Continues on the following page.

```
/*    --------------------------------------------------------
 *    | Arduino Experimentation Kit Example Code             |
 *    | CIRC-10 .: Temperature :. (TMP36 Temperature Sensor) |
 *    --------------------------------------------------------
 * A simple program to output the current temperature to the IDE's debug window
 * For more details on this circuit: http://tinyurl.com/c89tvd
 */


//TMP36 Pin Variables
int temperaturePin = 0; //the analog pin the TMP36's Vout (sense) pin is connected to
                        //the resolution is 10 mV / degree centigrade
                        //(500 mV offset) to make negative temperatures an option
double inputoriginal = 0.0;
double inputconverted = 0.0;

void setup()
{
  Serial.begin(9600);  //Start the serial connection with the computer
}

void loop()                // run over and over again
{
  //getting the voltage reading from the temperature sensor
  float temperature = getVoltage(temperaturePin);
  //converting from 10 mv per degree wit 500 mV offset
  // to degrees ((volatge - 500mV) times 100)
  temperature = (temperature - .5) * 100;
  Serial.print(temperature);
  Serial.print("   ");
  Serial.print(inputconverted);
  Serial.print("   ");
  Serial.print(inputoriginal);
  Serial.println(" "); //printing the result
  delay(1000);                          //waiting a second
}
```

```
/*
 * getVoltage() - returns the voltage on the analog input defined by
 * pin
 */
float getVoltage(int pin){
inputoriginal = analogRead(pin);
inputconverted = analogRead(pin) * .004882814;
return (analogRead(pin) * .004882814); //converting from a 0 to 1023 digital range
                // to 0 to 5 volts (each 1 reading equals ~ 5 millivolts
}
```

Example Code 5: Code for IR motion sensor.

```
//Code for detecting motion from an IRmotion sensor
//This is a digital only piece of hardware. Either there is motion or there isn't

//Declare Variables
int pirPin = 2; //digital 2 for input from motion sensor

void setup(){
 Serial.begin(9600);
 pinMode(pirPin, INPUT);
}

void loop(){
 int pirVal = digitalRead(pirPin);

 if(pirVal == LOW){ //was motion detected
  Serial.println("Motion Detected");
  delay(500);
 }
 if(pirVal == HIGH){ //was motion detected
  Serial.println("No Motion");
  delay(500);
 }


}
```

Example Code 6: Ping - For using the ultrasonic rangefinder (HC-04). If the Ultrasonic library has not already been installed, you must download the library and installed it as described.

```
/* YourDuino SKETCH UltraSonic Serial 1.0
 Runs HC-04 and hopefully SRF-06 Ultrasonic Modules
 Uses: Ultrasonic Library (Copy to Arduino Library folder)
 http://iteadstudio.com/store/images/produce/Sensor/HCSR04/Ultrasonic.rar
 terry@yourduino.com */

/*-----( Import needed libraries )-----*/
#include "Ultrasonic.h"
/*-----( Declare Constants and Pin Numbers )-----*/
#define  TRIG_PIN  13
#define  ECHO_PIN  12
/*-----( Declare objects )-----*/
Ultrasonic OurModule(TRIG_PIN, ECHO_PIN);
/*-----( Declare Variables )-----*/

void setup()   /****** SETUP: RUNS ONCE ******/
{
  Serial.begin(9600);
  Serial.println("UltraSonic Distance Measurement");
  Serial.println("YourDuino.com  terry@yourduino.com");

}//--(end setup )---

void loop()   /****** LOOP: RUNS CONSTANTLY ******/
{
  Serial.print(OurModule.Ranging(CM));
  Serial.print("cm   ");
  delay(100);  //Let echos from room dissipate
  Serial.print(OurModule.Ranging(INC));
  Serial.println("inches");

  delay(400);

}//--(end main loop )---
```

Example Code 7: Code for reading from a multiplexer. Coninued on the following page.

```
//Code for running through each of the pins 0-7 on a 8-way multiplexer (like a 4067).
//Reads all values from possible input and stores them in an array and prints to computer.

int analogPin = 0;    // Could be any analog pin
int val = 0;          // variable to store the value read
int row = 0;
double sec = 0.0;

//Array for storing values read from sensors connected to multiplexer
byte Values[]={0,0,0,0,0,0,0,0,0,0,0,0,0,0,0,0};

//Integers for storing bits. These bits become the HIGH or LOW values
//sent to the pins for activating the S0,S1,S2 pins on the multiplexter
int r0=0;
int r1=0;
int r2=0;

int i=0;    //Counter for loops


void setup()
{
  //Pins for activating S0,S1,S2 on the multiplexter. Can be any digital pins.
  pinMode(2, OUTPUT);
  pinMode(3, OUTPUT);
  pinMode(4, OUTPUT);

  Serial.begin(9600);         // setup serial
  Serial.println("CLEARDATA");  //For use with PLXDAQ
  Serial.println("LABEL,Time,Sec,Label1,Label2,Label3");
}
```

```
void loop()
{
 for (int i=0;i<=7;i++){  //Loop for writing to S0,S1,S2 on multiplexer
   r0=bitRead(i,0);
   r1=bitRead(i,1);
   r2=bitRead(i,2);
   digitalWrite(2,r0);
   digitalWrite(3,r1);
   digitalWrite(4,r2);
   delay(50);
   Values[i] = analogRead(analogPin);   // read the input pin
 }

 Serial.print("DATA,TIME,");
 sec=millis()/1000.0;
 Serial.print(sec);
 Serial.print(",");
 for (int i=0;i<=7;i++){   //Print all analog read values.
   Serial.print(Values[i]);
   Serial.print(",");
 }
 Serial.println(" ");
 row++;
 delay(10);
}
```

Example Code 8: Code for providing power to multiple devices through a multiplexer.

```
//Code for running through each of the pins 0-7 on a 8-way multiplexer (like a 4067).
//Provides power to all possible outputs.

int analogPin = 0;

//Integers for storing bits. These bits become the HIGH or LOW values
//sent to the pins for activating the S0,S1,S2 pins on the multiplexter
int r0=0;
int r1=0;
int r2=0;

int i=0;    //Counter for loops

void setup() {
  //Pins for activating S0,S1,S2 on the multiplexter. Can be any digital pins.
  pinMode(2, OUTPUT);
  pinMode(3, OUTPUT);
  pinMode(4, OUTPUT);

  Serial.begin(9600);         // setup serial
  Serial.println("CLEARDATA");  //For use with PLXDAQ
  Serial.println("LABEL,Time,Sec,Label1,Label2,Label3");
}

void loop() {
 for (int i=0;i<=7;i++){
   r0=bitRead(i,0);
   r1=bitRead(i,1);
   r2=bitRead(i,2);
   digitalWrite(2,r0);
   digitalWrite(3,r1);
   digitalWrite(4,r2);
   delay(50);
  }
}
```

Example Code 9: Code for the SPI_Master_Button. To be used with SPI_Slave_LED on next page.

```
//This code is for the master in the SPI pair with a slave. It communicates text from the master
//to the slave and the slave prints the text. This transfer occurs when the button is pressed
//on the master's circuit.

#include <SPI.h> //must include the spi library
int SSpin=43; //Slave Select pin. Used to tell the slave to listen to the message
int ButtonPin=22; //This pin provides power to the button
int ButtonRead=A15; //This pin detects the button press

void setup (void) {
  Serial.begin(9600);
  pinMode(SSpin, OUTPUT);
  pinMode(ButtonPin, OUTPUT);
  digitalWrite(SSpin, HIGH);  // ensure SS stays high for now
  digitalWrite(ButtonPin, HIGH);
  // SPI.begin() Puts SCK, MOSI, SS pins into output mode also puts SCK, MOSI into LOW state,
  // and SS into HIGH state. Then puts SPI hardware into Master mode and turn SPI on
  SPI.begin ();

  // Slow down the master a bit for less chance of lost data.
  SPI.setClockDivider(SPI_CLOCK_DIV8);
}

void loop (void) {
  if(analogRead(ButtonRead)<900){
    char c;
    digitalWrite(SSpin, LOW); // enable Slave Select
    // send test string. It is broken in to characters and sent individually
    for (const char * p = "Hello, world!\n" ; c = *p; p++){
      SPI.transfer (c);
    }
    digitalWrite(SSpin, HIGH); // disable Slave Select
    delay (500);
  }
}
```

Example Code 10: SPI_Slave_LED code for reading text message from SPI_Master_Button code and printing the message. Continues on the following page.

```
//This code is for the slave and paired with a master arduino through the SPI bus.
//When text is sent from the master, the slave captures each character and forms
//a full message which is then printed out.

#include <SPI.h>  //Must include the SPI library
int LEDpin=8; //pin to light the LED when the message is being recieved.
char buf [100]; //This buffer compiles the message into a coherent whole.
volatile byte pos;  //A counter that says what position in the buffer we are at.
volatile boolean process_it; //Tells when message is complete

void setup (void) {
  Serial.begin (9600);
  pinMode(LEDpin, OUTPUT);
  digitalWrite(LEDpin,LOW);
  pinMode(MISO, OUTPUT); //have to send on master in, *slave out*

  SPCR |= _BV(SPE); //turn on SPI in slave mode

  // get ready for an interrupt
  pos = 0;   // buffer empty
  process_it = false;

  SPI.attachInterrupt(); // now turn on interrupts
}
// main loop - wait for flag set in interrupt routine below
void loop (void) {
  if (process_it){
    digitalWrite(LEDpin, HIGH);
    buf [pos] = 0;
    Serial.println (buf);
    pos = 0;
    process_it = false;
  } // end of flag set
    delay(500);
    digitalWrite(LEDpin, LOW);
}
```

```
// SPI interrupt routine
ISR (SPI_STC_vect) {
 byte c = SPDR;  // grab byte from SPI Data Register

 if (pos < sizeof buf){  // add to buffer if room
  buf [pos++] = c;

  if (c == '\n'){ //newline means time to process buffer
   process_it = true;
  }
 }
}
```

Example Code 11: I2CMandS_Button2LED code for both the master and slave. Pressing a button on one will light an LED on the other Arduino. Continues on the following page.

```
/**
 *
 * Sample Multi Master I2C implementation.  Sends a button state over I2C to another
 * Arduino, which flashes an LED correspinding to button state.
 *
 * Connections: Arduino analog pins 4 and 5 are connected between the two Arduinos,
 * with a 1k pullup resistor connected to each line.  Connect a push button between
 * digital pin 10 and ground, and an LED (with a resistor) to digital pin 9.
 *
 */

#include <Wire.h>

#define LED 9
#define BUTTON 10

#define THIS_ADDRESS 0x8
#define OTHER_ADDRESS 0x9

boolean last_state = HIGH;

void setup() {
  Serial.begin(9600);
  pinMode(LED, OUTPUT);
  digitalWrite(LED, LOW);

  pinMode(BUTTON, INPUT);
  digitalWrite(BUTTON, HIGH);

  Wire.begin(THIS_ADDRESS);
  Wire.onReceive(receiveEvent);
  Serial.println("Setup ran");
}
```

```
void loop() {

  delay(12);
  Wire.beginTransmission(OTHER_ADDRESS);
  Wire.write(LOW);
  Serial.println(LOW);
  Wire.endTransmission();
  delay(12);
  Wire.beginTransmission(OTHER_ADDRESS);
  Wire.write(HIGH);
  Serial.println(HIGH);
  Wire.endTransmission();

  if (digitalRead(BUTTON) != last_state){
   last_state = digitalRead(BUTTON);
   Wire.beginTransmission(OTHER_ADDRESS);
   Wire.write(last_state);
   Serial.println(last_state);
   Wire.endTransmission();
  }
}

void receiveEvent(int howMany){
 while (Wire.available() > 0){
  boolean b = Wire.read();
  Serial.print(b, DEC);
  digitalWrite(LED, !b);
 }
 Serial.println();
}
```

Example Code 12: I2CMaster_Query_Recieve1 code for the master Arduino to request sensor data from a slave through I2C. Continued on the following page.

```
//Code for the master in a master slave pair for I2C.
//When a button is pressed on the master arduino, it queries the slave.
//The slave then sends sensor data back to the master which is printed.

#include <Wire.h>   //Must include the I2C library

#define LED 9
int BUTTON=10;

//Addresses for the master and slave. They must be swapped when uploading to the slave.
#define THIS_ADDRESS 0x8
#define OTHER_ADDRESS 0x9

int b=0;
int data[16];
boolean last_state = HIGH;

void setup() {
  Serial.begin(9600);
  pinMode(LED, OUTPUT);
  digitalWrite(LED, LOW);

  pinMode(BUTTON, INPUT);
  digitalWrite(BUTTON, HIGH);

  Wire.begin(THIS_ADDRESS);  //Starts the I2C communication between the arduinos
}
```

```
void loop() {
  if (digitalRead(BUTTON) == LOW){    //Is the button pressed?
   last_state = digitalRead(BUTTON);
   Serial.println(last_state);
   //Sends a 1 to the slave as a command to sample the sensor.
   Wire.beginTransmission(OTHER_ADDRESS);
   Wire.write(1);
   Wire.endTransmission();
   //*****************************************
   delay(1000);   //Important delay to give slave time to sample sensor.
   requestData();  //function for requesting data from the slave
  }
}

void requestData(){
 int count=Wire.requestFrom(OTHER_ADDRESS, 16);  //Requests 16 bytes of data from slave
 if(count==16){       //If data is appropriate size, read the data.
  for(int i=0;i<16;i++){  //Reads the data one byte at a time.
   b=Wire.read();
   data[i]=b;        //Stores the bytes.
  }
  Serial.println();
  for(int i=0;i<8;i++){            //Convert data to temperature.
   float temperature = data[i]+(data[i+8]/100.0);
   Serial.print(temperature);
   Serial.println();
  }
 }
 else{            //If data was not appropriate size.
  Serial.println("Data was not received");
 }
}
```

Example Code 13: I2CSlave_TempRead1 code for accepting commands from a master Arduino and sending back sensor data from a temperature sensor. Continued on the following two pages.

```
//For a slave in a master slave pair of arduinos.
//The master requests sensor data from the slave. This slave reads from a temperature
//sensor and then sends the data back to the master

#include <Wire.h>  //Must include the I2C library
#define LED 9
//Addresses for the Slave and Master.
#define THIS_ADDRESS 0x9
#define OTHER_ADDRESS 0x8
int b=0;
byte val[]={16};  //Array for sending the first 16 bytes of data to the Master Arduino

float temperature = 0.0;
float temperatureC = 0.0;
float temperatureF = 0.0;
int temperatureFfirst = 0;
int temperatureFsecond = 0;
#define temperaturePin A0 //Just like in the previous Temp Meas sketch

int inputoriginal = 0.0;
double inputconverted = 0.0;

void setup() {
  Serial.begin(9600);       // start serial for output
  pinMode(LED, OUTPUT);
  digitalWrite(LED, LOW);

  Wire.begin(THIS_ADDRESS);   //Begins the I2C communication between the arduinos
  Wire.onReceive(receiveEvent);  //Defines a function that interrupts when data is received
  Wire.onRequest(requestEvent);  //Defines a function that interrupts when a request is
                                 //received
}
```

```
void loop() {
  if(b==1){
    digitalWrite(LED,HIGH);   //Turns on LED
    for(int i=0;i<8;i++){
      TempMeas();
      val[i]=temperatureFfirst;
      val[i+8]=temperatureFsecond;
    }
    b=0;
  }
}
//*****************************************************************
//*****************************************************************
//function that runs when Master Sends data to this Arduino
void receiveEvent(int howMany){
  while(Wire.available()){
    b = Wire.read();     //While data is comming in, read values into b
  }


}
//*****************************************************************
//function that runs when Master sends request for data
void requestEvent(){
  delay(20);  //Wait a bit, not needed?
  Wire.write (val,16);  //Send 16 bytes of data to master
  digitalWrite(LED,LOW);    //Turn off LED
}
```

```
//****************************************************************
//function that measures the temperature from the sensor
void TempMeas(){
  delay(10);
  inputoriginal = 0.0;
  temperature = 0.0;
  temperatureF = 0.0;
  temperatureFfirst = 0;
  temperatureFsecond = 0;
  inputoriginal = analogRead(temperaturePin);
  temperature = inputoriginal * .004882814;
  temperatureF = (temperature - .5) * 100;   //converting from 10 mv per degree wit 500 mV offset
  temperatureFfirst = temperatureF;
  temperatureFsecond = 100*(temperatureF - temperatureFfirst);
  //These serial prints are not necessary, but good to see if data is correct before transmission.
    Serial.print(inputoriginal);
    Serial.print(",");
    Serial.print(temperatureF);
    Serial.print(",");
    Serial.print(temperatureFfirst);
    Serial.print(",");
    Serial.println(temperatureFsecond);

}
```

Example Code 14: MCP9808_SoftI2C sketch for reading from the sensor with a software I2C bus. Continues on the following 6 pages.

```
//Sketch to read the MCP9808 using the SoftI2CMaster library. Adapted from BMA example.
//This library has many advantages over the hardware version, including ability to make
//multiple I2C buses on the same Arduino, embedded timeout functions to prevent a less than
//perfect bus from freezing when a packet is dropped, and ability to alter transmission speeds.

// use low processor speed (you have to change the baud rate to 2400!)
// #define I2C_CPUFREQ2 (F_CPU/8)
#define NO_INTERRUPT 0    //?
#define I2C_TIMEOUT 100   //wait 100ms for a response. If none, then break and restart
#define I2C_SLOWMODE 1    //operates at 25kHz no matter the CPU speed
#define FAC 1             //variable to use on the next line
#define I2C_CPUFREQ (F_CPU/FAC)  //ability to scale the CPU speed used by I2C


int a = 0;

int I2Cpin1 = 22;
int I2Cpin2 = 23;

//Pins you use for I2C (SDA and SCL) must be defined using the port and pin number
#define SDA_PORT PORTC  //includes pins 30-37
#define SDA_PIN 7       //pin 30
#define SCL_PORT PORTC  //includes pins 30-37
#define SCL_PIN 6       //pin 31

#include <SoftI2CMaster.h>  //Must include the I2C software library
#include <avr/io.h>         //Required for I2C soft

#define BMAADDR 0x30        //Address for the MCP9808

int xval, yval, zval;
float zval1;
```

```
/*
 * Example of how to add a second I2C bus to the same board.
 * In addition to the following lines, the library must be altered as seen in
 * SoftI2CMaster2.h, each variable must be altered because the variable names
 * cannot be used in two different libraries. I did this by adding a '2' to
 * each name. This can be done as many times as you have pins available for buses.
#define NO_INTERRUPT2 0
#define I2C_TIMEOUT2 100
#define I2C_SLOWMODE2 1
#define FAC2 1
#define I2C_CPUFREQ2 (F_CPU/FAC2)

#define SDA_PORT2 PORTC
#define SDA_PIN2 5
#define SCL_PORT2 PORTC
#define SCL_PIN2 4

#include <SoftI2CMaster2.h>
*/
//***********************************************************
//Slows down the processor. Not sure if needed.
void CPUSlowDown(int fac2) {
  // slow down processor by a fac2
   CLKPR = _BV(CLKPCE);
   CLKPR = _BV(CLKPS1) | _BV(CLKPS0);
}
```

```
//************************************************************
//Writes initialization commands to the MCP9808, telling it that
//we will be sampling tempterature as normal.
boolean setControlBits(uint8_t cntr) {
  Serial.println(F("Soft reset"));
  if (!i2c_start(BMAADDR | I2C_WRITE)) {  //if it cannot communicate, false
    return false;
  }
  if (!i2c_write(0x0A)) {  //If it could not write command, false
    return false;
  }
  if (!i2c_write(cntr)) {  //if it could not write command, false
    return false;
  }
  i2c_stop();      //stops i2c for the time being
  return true;
}
//************************************************************
//Initialize MCP9808. Calls setControlBits above
boolean initBma(void) {
  if (!setControlBits(B00000010)) return false;;
  delay(100);
  return true;
}
//************************************************************
//read a value from the MCP9808 in the form of two 8bit bytes
//The two are then combined into a 16bit byte.
int readOneVal(boolean last) {
  uint16_t msb, lsb;       //define two unsigned ints; msb and lsb
  msb = i2c_read(false);   //store first 8bits into msb
  msb <<= 8;               //shift those 8 bits to beginning of msb
  msb |= i2c_read(last);   //read last 8bits and concatanate with first 8bits
  if (last) i2c_stop();    //if reached end of values, stop I2C
  return (int)(msb);       //return the 16bit value
}
```

```
/*
//***********************************************************
//If a second i2c bus is used, this will read from it, same as above
int readOneVal2(boolean last) {
  uint16_t msb, lsb;
  msb = i2c_read2(false);
  msb <<= 8;
  msb |= i2c_read2(last);
  if (last) i2c_stop2();
  return (int)(msb);
}
*/
//***********************************************************
//Reads values from MCP9808 via I2C and stores values
boolean readBma(void) {
  zval = 0xFFFF;        //variable for storing bytes from MCP9808
  zval1 = 0xFFFF;       //variable for conversion of byte to temperature
  //If I2C cannot communicate with sensor, return false
  if (!i2c_start(BMAADDR | I2C_WRITE)) return false;
  if (!i2c_write(0x05)) return false;
  if (!i2c_rep_start(BMAADDR | I2C_READ)) return false;
  zval = readOneVal(true);  //calls readOneVal above
  zval1 = zval & 0x0FFF;    //bitwise operation for conversion to temperature
  zval1 = zval1/16.0;       //final conversion operation
  return true;
}
```

```
/*
//*************************************************************
//Reads values from second I2C bus if used. same as above
boolean readBma2(void) {
 zval = 0xFFFF;
 zval1 = 0xFFFF;
 if (!i2c_start2(BMAADDR | I2C_WRITE2)) return false;
 if (!i2c_write2(0x05)) return false;
 if (!i2c_rep_start2(BMAADDR | I2C_READ2)) return false;
 zval = readOneVal2(true);
 zval1 = zval & 0x0FFF;
 zval1 = zval1/16.0;
 return true;
}
*/
//*************************************************************
//Setup
void setup(void) {
 Serial.begin(9600); // in case of CPU slow down, change to baud rate / 8!
 #if FAC != 1
  CPUSlowDown(FAC);
 #endif
 Serial.println(F("Intializing 0..."));
 Serial.print("I2C0 delay counter: ");
 Serial.println(I2C_DELAY_COUNTER);
 if (!i2c_init())
  Serial.println(F("Initialization error0. SDA or SCL are low"));
 else
  Serial.println(F("...done0"));
```

```
/*
//needed only if using a second I2C bus
#if FAC2 != 1
  CPUSlowDown(FAC2);
#endif
Serial.println(F("Intializing2 ..."));
Serial.print("I2C delay counter2: ");
Serial.println(I2C_DELAY_COUNTER2);
if (!i2c_init2())
  Serial.println(F("Initialization error2. SDA or SCL are low"));
else
  Serial.println(F("...done2"));
*/
//Needed?
pinMode(I2Cpin1,OUTPUT);
pinMode(I2Cpin2,OUTPUT);
digitalWrite(I2Cpin1, HIGH);
digitalWrite(I2Cpin2, LOW);
//?
}
```

```
//*********************************************************
//Main loop
void loop(void){
  digitalWrite(I2Cpin1, HIGH);   //Needed?
  digitalWrite(I2Cpin2, LOW);    //Needed?
  a++;                //Needed?

  if (!readBma()) Serial.println(F("READ ERROR0"));  //If read fails,
  //Print the values
  Serial.print("time= ");
  float time=millis()/1000.0;
  Serial.print(time);
  Serial.print("   ");
  Serial.print(F(" Z1="));
  Serial.println(zval1);
  delay(150);

  /* Needed only if using a second I2C bus
  digitalWrite(I2Cpin2, HIGH);   //Needed?
  digitalWrite(I2Cpin1, LOW);    //Needed?

  if (!readBma2()) Serial.println(F("READ ERROR2"));
  Serial.print("time= ");
  time=millis()/1000.0;
  Serial.print(time);
  Serial.print("   ");
  Serial.print(F(" Z2="));
  Serial.println(zval1);
  delay(150);
  */
}
```

Index

Amplifier
    AC, 72, 75–81
    DC, 73–75
    Inverting, 91
    Non-Inverting, 89–90
Analog to Digital Conversion (ADC), 109–10
Arduino, 102–57
Breadboard, 16
Capacitor
    Ceramic Codes, 14
    Theory, 20
Digital Multimeter (DMM), 13
Diode
    I-V Curve, 43–47
    Signal, 56
    Theory, 41–43
    Zener, 43, 47
Inductor, 31
Integrating Circuit, 92–97
Inter-Integrated Circuit (I2C), 146–50
LR Circuit, 33–34
LRC Circuit, 35–36, 39
Motor
    Brushless, 120
    D.C., 117
    Stepper, 117
Multiplexer, 137, 155–57, 145
Operational Amplifier (OpAmp)
    Slew Rate, 86–87
    Theory, 84–85
    Unity Gain Buffer, 88
Oscilloscope, 9–12, 21–23
    Probes, 10
Passive Filters, 27
    Bandpass, 28
    High Pass, 78
    Low Pass, 27
    Multi-stage, 29
Piezo Buzzers, 123
PLX-DAQ, 110
Potentiometer, 64, 76, 109, 123
Processing, 123
Pulse Width Modulation (PWM), 56
RC Circuits, 24–28
Rectifier
    Full Wave, 51–55
    Half Wave, 48–50
Resistor Color Codes, 15
Sensor
    Accelerometer, 132
    Infrared Motion, 130
    Photo-resistor, 128
    Temperature, 127
    Ultrasonic Rangefinder, 133
Seven-Segment Display, 114
Shield
    Motor Control, 116–18
    Relay, 119
    SD Card, 121
Speakers, 123
Transformer, 37–38, 40
Transistor
    Common Base, 65
    Common Collector, 67
    Common Emitter, 69
    Darlington Pair, 82
    Theory, 60–63
Voltage Divider, 16, 17